The Evolution of Sustainable Investments and Finance

Francesco Gangi · Antonio Meles ·
Lucia Michela Daniele · Nicola Varrone ·
Dario Salerno

The Evolution of Sustainable Investments and Finance

Theoretical Perspectives and New Challenges

palgrave
macmillan

Francesco Gangi
Department of Economics
University of Campania Luigi Vanvitelli
Capua, Italy

Antonio Meles
Department of Economics
University of Campania Luigi Vanvitelli
Capua, Italy

Lucia Michela Daniele
Department of Economics
University of Campania Luigi Vanvitelli
Capua, Italy

Nicola Varrone
Department of Economics
University of Campania Luigi Vanvitelli
Capua, Italy

Dario Salerno
Department of Business and Quantitative Studies
Parthenope University of Naples
Naples, Italy

ISBN 978-3-030-70349-3 ISBN 978-3-030-70350-9 (eBook)
https://doi.org/10.1007/978-3-030-70350-9

This Palgrave Macmillan imprint is published by the registered company Springer Nature Switzerland AG
The registered company address is: Gewerbestrasse 11, 6330 Cham, Switzerland

Acknowledgements

We would like to thank Palgrave for believing in this project and for supporting it from its first stages. Special thanks go to Hemapriya Eswanth, Project Coordinator, and Lucy Kidwell, Assistant Editor. Thanks also to Dean Bargh for the thorough language edit. Finally, we would like to thank the reviewers for their helpful advice and Maria Coscia (Ph.D. Student) for her contribution to the bibliography research and systematization. Of course, the content and the quality of this book remains the sole responsibility of the authors.

Introduction

The concept of socially and environmentally responsible investment has been circulating among academics and professionals for some years now. However, since the first attempts were made at defining the socially responsible investor (e.g. Cowton 1999; Michelson et al. 2004), much has happened, not least the severe financial crisis of 2007–2008, serving as a crowning example of an ethical crisis and its woeful consequences (US Financial Crisis Commission 2011). Concurrently, while this drama played out on economies worldwide, interest has been growing among academics and practitioners in financial models that have a positive impact on communities' well-being and that safeguard the natural environment.

This book's aim is to provide an evolutionary analysis of sustainable investment and finance, with all the chapters sharing an evolutionary logic in their approach to the topic. They are linked by a common denominator, which is the increasing attention investors, firms, financial institutions and policymakers are now paying to the topic of social, environmental and corporate governance responsibilities. As such, the book provides a unifying message and approach, demonstrating how this topic is cutting across boundaries, describing the behaviours of private investors, the new wave of socially responsible banking, the involvement of communities in investment activities, and the new rules of equity finance and asset allocation.

The evolutionary methodology will help us transcend the age-old and outdated trade-off argument in which better social outcomes are at the expense of lower financial gains. There is an urgent need for a radical rethink about the "separation" model in finance, as testified by the new millennium's grave social crises caused by irresponsible financial models for the community and within businesses, as well as the increasingly frequent environmental calamities caused by climate change and corporate behaviour. More specifically, the book will lay to rest the anachronistic false duality between the shareholder view (Friedman 1970; Jensen and Meckling 1976; Jensen 2002) and the stakeholder view (Freeman 1984; Donaldson and Dunfee 1994; Donaldson and Preston 1995), which for too many years has lit the fire beneath the self-limiting question: *does it pay to be good?* (Barnett and Salomon 2012).

Social and environmental sustainability is no longer an optional extra, an add-on or a bonus. Despite previous voices of caution (Juravle and Lewis 2008), socially responsible investing (SRI)'s transition from niche to mainstream presents a path paved with new opportunities for investors and firms. Such an evolution is encouraging research and professional trajectories that are intercepting the logic of sustainability at multiple levels. The current book aims to address these issues from new and different angles.

In analysing the evolution of SRI from niche to the mainstream of financial markets, we will find ourselves taking a fresh look at sustainability as an opportunity for innovative business and financial models. A second point is that a focus on environmental, social and governance (ESG) issues in alternative investments and in the banking industry is consistent with an appropriate response to the considerable challenges presented by the financial crises, which is related to the need to recover the original social function of banks and backers (Schumpeter 1934). Furthermore, the emergence of digital finance has opened up new ways of financing businesses and social projects through the involvement of the "crowd". At the same time, the development of entrepreneurship requires new models, including venture capital activities, which, in assessing the viability of a project, cannot rate its economic sustainability while ignoring its social sustainability.

This book therefore aims to fully investigate all the above topics, and in doing so fill a gap in the literature by presenting a comprehensive interpretation of sustainable investment and financing models including firms and funders. This objective will be supported by theoretical reflections and empirical descriptions.

The book's five chapters are organised with the aim of conveying the desired sense of an evolutionary trajectory, with a circular linkage among them still being apparent. They are as follows:

Chapter 1: Socially Responsible Investment (Sri): From Niche to Mainstream
Chapter 2: Esg Criteria in Alternative Investments
Chapter 3: Socially Responsible Banking: Towards a New Firm–Bank Relationship
Chapter 4: Crowdfunding and Socially Responsible Initiatives
Chapter 5: Venture Capital and Responsible Investing: Progress, Problems and Perspectives

Chapter 1 begins by providing an overview of ethical and socially responsible finance, presented as a new paradigm that transcends the separation model between social concerns and financial returns. In this regard, SRI represents a form of investment capable of meeting the expectations of investors interested in both profit and social responsibility. The chapter goes on to analyse socially responsible funds (SRFs) from different viewpoints. First, it focuses on SRFs' capital allocation processes through the nexus agency model. Second, it discusses the prior literature on financial performance in relation to social responsibility screens, then, third, the opportunities and risks associated with the transition of SRI from a niche to the mainstream of financial markets. Fourth, the chapter analyses the characteristics of the ESG ratings of investment mutual funds. A comparison between the European and US SRI markets concludes the chapter.

Chapter 2 looks more deeply at ESG criteria in alternative investments. Traditionally, ESG investments have been oriented towards liquid assets such as securities and bonds, due to the disclosure of information, the weight of assets under management, and rights provided to shareholders. Nowadays, ESG aspects are properly incorporated into alternative asset classes to a significant extent, thanks to their association with risk reduction and value creation. The purpose of this chapter is to explore the nexus between ESG criteria and alternative investments, by shedding light on this trend's past and present and predicting its future evolution. The first section provides an overview of alternative investments.

The second examines the relationship between ESG factors and alternative investments. The third section discusses ESG investment policy in the private equity (PE) industry. Unlike other categories of alternative investment, PE investors usually apply some ESG criteria as filtering factors for their investment opportunities during the screening process; the fourth section therefore follows with a focus on the ESG due diligence process in selecting PE-backed firms. The fifth section zooms in on ESG factors in hedge fund strategies. The speculative nature of hedge funds may seem at odds with a socially responsible and environmentally oriented policy, but ESG factors are often linked directly with notions of investor activism and good governance. The final section focuses on ESG in real estate investments.

Chapter 3 advances premises and unravels the implications of a new mode of financial decision-making, with specific reference to the relationship between the firm and the bank. In recent years, the banking industry has begun to pay more serious attention to its role as a global system in service to society. The purpose of this chapter is therefore to analyse how and why banks are engaging in socially responsible initiatives, again described from an evolutionary perspective. The first part draws from the "Schumpeterian view" of economic development and introduces the social role of the banking system in support of economic growth and sustainable prosperity. With these premises established, the second part analyses the growing phenomenon of the integration by banks of ESG principles into their business practices. Special emphasis is placed on the inclusion of ESG factors within bank lending processes. Indeed, it is well accepted in the literature (Simpson and Kohers 2002; Thomson and Cowton 2004; Weber 2012) that not only do banks have direct responsibilities towards society, but they also bear an indirect responsibility as a function of the level of social and environmental engagement of their borrowers. Indeed, the banking system has attracted frequent criticism for its inability to take into account the impact on society of its strategies and operations (e.g. Cornett et al. 2016). Against this backdrop, banks are being called on to demonstrate the fulfilment of their social obligations (Menassa and Brodhacker 2017) and, as such, engagement in ESG initiatives represents for banks one of the most valuable ways in which to recover lost reputation and improve relationships with stakeholders (e.g. Ruiz et al. 2014; Cornett et al. 2016). Accordingly, in the last section, the chapter makes a comparison between the financial results of so-called ethical banks and those of conventional ones.

Chapter 4 looks at the evolution of alternative financial models, and their contribution to sustainable investment and finance. The Internet, and to an even larger degree digital sharing technologies, represent a strong force for discontinuity in the relationship between financial institutions and entrepreneurs. Alternative financial models, such as crowdfunding, have spread rapidly worldwide as innovative ways of financing both businesses and social projects. By democratising the access to financial resources, crowdfunding dramatically enlarges the audience of potential supporters of new projects. The chapter's structure is as follows: the first section introduces the alternative finance landscape, crowdfunding mechanisms and the main actors, focusing on crowdfunding models and on the role of the crowd. The second section explores the nexus between crowdfunding models and entrepreneurship, while the third frames crowdfunding as a value co-creation process. Crowdfunding, in addition, is offering new opportunities for socially responsible initiatives, and the fourth section, accordingly, discusses how crowdsourcing can provide socially oriented projects and organisations with adequate financial resources.

Chapter 5, the final chapter, analyses a recent, topical phenomenon: the link between venture financing and socially responsible investing. Over the past few decades, the global venture capital (VC) industry has mushroomed mainly because of the widely recognised capacity of VC firms to create value for their investors (e.g. Gompers and Lerner 1999; Megginson et al. 2019). However, these skills seem to be no longer sufficient to drive further expansion of the industry, and VC firms are going to be facing some new and important challenges in the coming years. Accordingly, numerous limited partners (LPs) of VC funds around the world are now looking at ESG goals as a priority. Of course, behind this radical change in VC firms' investment philosophy lies more than just philanthropy: there is also a belief that ESG criteria might help them "do good while doing well". Numerous studies recognise that firms that invest in ESG do not have to accept a decline in performance as a consequence (e.g. Jo and Harjoto 2012; Awaysheh et al. 2020; Gangi et al. 2020). The second challenge that VCs have been called on to embrace is that of gender in entrepreneurship (e.g. Guzman and Kacperczykb 2019; Balachandra 2020). In the US, for example, only 3% of VC funds are channelled into financing all-female teams, with 8% going to mixed teams and the remaining 89% to all-male teams (Crunchbase News 2019). The lack of female representation is evident not only among start-up founders but also inside VC firms. According to the recently published *New York Times* and CBInsights 2018 list, only nine out of the 100 top VC partners worldwide are women. Such a lack of gender representation is unacceptable from a social point of view and carries with it a significant opportunity cost for both individual companies and the entire economy. Indeed, past research has documented that firms with greater gender diversity are considered more creative, more innovative and more profitable (e.g. Campbell and Mínguez-Vera 2008; Ruiz-Jiménez et al. 2016). Against this backdrop, the aim of the fifth chapter is first to analyse the key characteristics of VC investments and the degree of implementation of ESG criteria within the VC industry. Second, it discusses how ESG compliance might help VC firms to raise funds, select and build successful firms, and facilitate their divestments. Finally, the chapter enquires into the characteristics and perspectives of the social VC and the gender gap in entrepreneurship and the VC industry.

We hope that this book will appeal to academics, entrepreneurs and professionals involved in the complex world of investing and finance. It is a world that is in the midst of major transformations, among which ethical and socially responsible approaches represent one of the main challenges of the new millennium for firms and investors.

Francesco Gangi
Antonio Meles
Lucia Michela Daniele
Nicola Varrone
Dario Salerno

References

Awaysheh A, Heron RA, Perry T, Wilson JI (2020) On the relation between corporate social responsibility and financial performance. Strateg Manag J 41(6):965–987

Balachandra L (2020) How gender biases drive venture capital decision-making: exploring the gender funding gap. Gend Manag

Barnett ML, Salomon RM (2006) Beyond dichotomy: the curvilinear relationship between social responsibility and financial performance. Strateg Manag J 27 (11):1101–1122. https://doi.org/10.1002/smj.557

Campbell K, Mínguez-Vera A (2008) Gender diversity in the boardroom and firm financial performance. J Bus Ethics 83(3):435–451

Cornett MM, Erhemjamts O, Tehranian H (2016) Greed or good deeds: An examination of the relation between corporate social responsibility and the financial performance of US commercial banks around the financial crisis. J Bank Financ 70:137–159. https://doi.org/10.1016/j.jbankfin.2016.04.024

Cowton C (1999) Playing by the rules: ethical criteria at an ethical investment fund. Bus Ethics 8(1):60–69. https://doi.org/10.1111/1467-8608.00126

Crunchbase News (2019) Inside VC firms: the gender divide, August 14, 2019

Donaldson T, Preston LE (1995) The stakeholder theory of the corporation: concepts, evidence, and implications. Acad Manage Rev 20(1):65–91. https://doi.org/10.5465/amr.1995.9503271992

Freeman RE (1984) Strategic Management: a Stakeholder Approach. Pitman, Boston

Friedman M (1970) The social responsibility of business is to increase its profits. The New York Times Magazine (September 13) (Retried on 13 January 2012 from: http:// www.colorado.edu/studentgroups/libertarians/issues/fri edman-soc-resp-business. html)

Gangi F, Daniele LM, Varrone N (2020) How do corporate environmental policy and corporate reputation affect risk-adjusted financial performance? Bus Strategy Environ 29(5):1975–1991. https://doi.org/10.1002/bse.2482

Gompers PA, Lerner J (1999) What drives venture capital fundraising? (No. w6906). National bureau of economic research

Guzman J, Kacperczyk AO (2019) Gender gap in entrepreneurship. Res Policy 48(7):1666–1680

Jensen M (2002) Value maximization, stakeholder theory, and the corporate objective function. Bus Ethics Q 235–256. https://doi.org/10.2307/385 7812

Jensen M, Meckling W (1976) Theory of the firm: managerial behavior, agency costs, and ownership structure. J Financ Econ 3:305–383. https://doi.org/ 10.1016/0304-405x(76)90026-x

Jo H, Harjoto MA (2012) The causal effect of corporate governance on corporate social responsibility. J Bus Ethics 106(1):53–72. https://doi.org/10.1007/ s10551-011-1052-1

Juravle C, Lewis A (2008) Identifying impediments to SRI in Europe: a review of the practitioner and academic literature. Bus Ethics Eur Rev 17(3):285–310. https://doi.org/10.1111/j.1467-8608.2008.00536.x

Megginson WL, Meles A, Sampagnaro G, Verdoliva V (2019) Financial distress risk in initial public offerings: how much do venture capitalists matter? J Corp Finance 59:10–30

Michelson G, Wailes N, van der Laan S, Frost G (2004) Ethical investment processes and outcomes. J Bus Ethics 52(1):1–10. https://doi.org/10.1023/ b:busi.0000033103.12560.be

Ruiz-Jiménez JM, del Mar Fuentes-Fuentes M, Ruiz-Arroyo M (2016) Knowledge combination capability and innovation: the effects of gender diversity on top management teams in technology-based firms. J Bus Ethics 135(3):503–515

Schumpeter JA (1934) The theory of economic development. Harvard: Economic Studies

Simpson WG, Kohers T (2002) The link between corporate social and financial performance: evidence from the banking industry. J Bus Soc Rev 35(2):97–109. https://doi.org/10.1023/a:1013082525900

Weber O (2012). Environmental credit risk management in banks and financial service institutions. Bus Strategy Environ 21(4):248–263. https://doi.org/10.1002/bse.737

Contents

1 Socially Responsible Investment (SRI): From Niche to Mainstream 1
Francesco Gangi, Nicola Varrone, and Lucia Michela Daniele
1.1 Beyond the Separation Model 1
1.1.1 The Separation Model and the Growth of Inequality and Imperfections 3
1.1.2 The Separation Model Within Firm Theory: Profit Maximisation Is not a Convincing End-Goal 7
1.2 The Main Features of Ethical and Socially Responsible Investing 11
1.2.1 Capital Allocation by Socially Responsible Funds 13
1.2.2 Decades of Studies on SRI Performance 17
1.2.3 The Drivers and Characteristics of the Individual Socially Responsible Investor 21
1.3 The Fiduciary Duties of SRFs: The Nexus Agency Model 24
1.4 The Opportunities and Risks of SRI as a Mainstream 27
1.5 ESG Rating of Investment Funds 29
1.6 Institutional Recognition of SRI: EU Regulatory Norms 33

1.7 SRI Market Trends: An EU and US Comparison 36
1.7.1 The US SRI Market 42
References 46

2 ESG Criteria in Alternative Investments 59
Dario Salerno
2.1 Introduction 59
2.2 Alternative Investments: An Overview 60
2.3 ESG Factors and Alternative Investments 63
2.3.1 The Role of Impact Investment 70
2.3.2 Vehicles for Impact Investments 72
2.4 Private Equity 74
2.4.1 ESG Selection Criteria in Private Equity 75
2.4.2 Impact Investment Criteria in Private Equity 79
2.5 PE Investments and ESG Due Diligence 81
2.6 ESG Aspects in Hedge Fund Investments 85
2.7 Environmental and Sustainable Real Estate Investments 87
2.7.1 Sustainable Activity in the Real Estate Industry 90
References 92

3 Socially Responsible Banking: Towards a New Firm–Bank Relationship 101
Francesco Gangi, Nicola Varrone, and Lucia Michela Daniele
3.1 A Schumpeterian View: The Social Role of Bank 101
3.1.1 The Evolution of Bank Interventions in "Bank-Centred" Systems: The Italian Case 105
3.2 Why Do Banks Comply with ESG Principles? 108
3.2.1 ESG Principles in Banking as Drivers of Business Benefits 113
3.3 An Overview of ESG Criteria in the Banking Sector 117
3.4 ESG Investment Policy and Bank Lending 121
3.5 ESG and Potential Impacts on Banks' Performance 125
3.6 Banks' Social Disclosure 128
3.6.1 BSD Relative to the ESG Pillars 131
3.6.2 BSD and Stakeholder Categories 135
3.7 Ethical Versus Conventional Banks: An Explorative Comparative Analysis 137
References 144

4 **Crowdfunding and Socially Responsible Initiatives** 155
Lucia Michela Daniele
4.1 *Fintech and the Alternative Financial Landscape: An Introduction to Crowdfunding* 155
4.1.1 *Community and Financially Oriented Crowdfunding Models* 158
4.1.2 *Signals in Crowdfunding: The Role of the Crowd* 160
4.2 *The Relevance of Crowdfunding in Entrepreneurial Finance* 166
4.3 *The Crowdfunding Ecosystem for Innovation and Value Co-creation* 171
4.4 *Crowdfunding and Sustainability* 176
4.4.1 *Crowdfunding and CSR* 177
4.4.2 *CF as a Form of Impact Finance* 191
4.4.3 *Sustainability-Oriented CF Projects and Crowd Behaviour* 193
References 196

5 **Venture Capital and Responsible Investing: Progress, Problems and Perspectives** 211
Antonio Meles
5.1 *Introduction* 211
5.2 *The Key Characteristics and the Main Stages of VC Investments* 212
5.3 *ESG Implementation in the VC Industry: Motivation, Diffusion and Strategies* 218
5.4 *How Does ESG Compliance Contribute to VCs' Success?* 223
5.4.1 *ESG and VC Firms' Fundraising* 223
5.4.2 *Role of ESG in Mitigating Problems of Information Asymmetry Faced by VC Firms* 226
5.4.3 *Are ESG-Compliant VC Firms More Likely to Add Value to Investee Firms?* 229
5.4.4 *ESG Considerations at Each Stage of the Investment Process: Some Operational Indications* 231
5.5 *When a VC Becomes a Social VC* 233
5.6 *The Dark Side of VC: Addressing the VC Gender Gap* 241
References 244

Conclusion 251

Index 257

List of Figures

Fig. 1.1 Comparison of Regulation Index (*Source* Fraser Institute) 5
Fig. 1.2 EU SRI total market volume, 2010–2018 (*Source* Authors' elaboration of GSIA [Global Sustainable Investment Alliance] Global Sustainable Investment reports. Reports consulted: 2012, 2014, 2016 and 2018) 37
Fig. 1.3 Total EU SRI investment strategies by yearly market share (*Source* Authors' elaboration of data gathered from Eurosif biennial SRI market reports. Reports consulted: 2012, 2014, 2016 and 2018) 41
Fig. 1.4 EU SRI: diffusion in the Eurozone, 2018 (*Source* Authors' elaboration of GSIA report 2018. European SRI Study 2018) 41
Fig. 1.5 US SRI, total market volume, 2010–2018 (biennial CAGR) (*Source* Authors' elaboration of GSIA [Global Sustainable Investment Alliance] Global Sustainable Investment reports. Reports consulted: 2010–2012, 2012–2014, 2014–2016 and 2016–2018) 44
Fig. 1.6 US SRI ESG strategies: growth rate, 2016–2018 (*Source* authors' elaboration of Forum for Sustainable and Responsible Investment biennial reports. Reports consulted: 2016 and 2018) 45
Fig. 1.7 EU and US SRI market comparison, 2010–2018 (*Source* authors' elaboration of GSIA [Global Sustainable Investment Alliance] Global Sustainable Investment reports. Reports consulted: 2012, 2014, 2016 and 2018) 45

Fig. 1.8 EU and US comparison: average growth rate by investment strategy, 2010–2018 (*Source* Authors' elaboration of USSIF [The Forum for Sustainable and Responsible Investment] and GSIA reports. Reports USSIF and GSIA consulted: 2012, 2014, 2016 and 2018) 46
Fig. 2.1 The impact investment journey (*Source* Cambridge Associates [2015] and New Philanthropic Capital [2015]) 71
Fig. 2.2 The spectrum of capital (*Source* Nicklin [2012], Clara Barby, Bridges Ventures) 73
Fig. 3.1 The UNEP FI principles for responsible banking (*Source* www.unepfi.org) 111
Fig. 4.1 Global onboarding and success rate of CF models (*Source* Authors' elaboration of CCAF data, April 2020, in *The Global Alternative Finance Market Benchmarking Report*. Onboarding rate: the rate at which CF proponents are allowed to proceed with a fundraising campaign. Successful funding: percentage of onboarded firms that effectively collected capital from the platform) 167
Fig. 5.1 Main entities involved in a VC fund (*Source* Adapted from EVCA) 215
Fig. 5.2 European VC funds' motivations for ESG engagement (*Source* Adapted from Botsari and Lang [2020]) 221
Fig. 5.3 ESG strategies of European VC funds (*Source of data* Botsari and Lang [2020]) 222

List of Tables

Table 1.1	SRI investment classification: EU and US	39
Table 1.2	Total EU SRI investment strategies by year	40
Table 1.3	EU SRI: total AUM by region and strategy, 2010–2018	43
Table 1.4	US SRI: investment strategy by market share, 2010–2018	44
Table 2.1	Alternative investments categories	62
Table 2.2	Overview of each of the three value creation streams	67
Table 3.1	Italian GDP and financial debts of firms	106
Table 3.2	Diffusion of voluntary principles for socially responsible banking	113
Table 3.3	An analysis of the BSD of leading commercial and systemic banks in the EU, by ESG pillar	132
Table 3.4	An analysis of the BSD of leading commercial and systemic banks in the EU, by stakeholder category	133
Table 3.5	Institutions included in the study	141
Table 3.6	Description of variables	142
Table 3.7	Credit quality of ethical and conventional banks: univariate analysis	142
Table 4.1	CF and CSR: CF projects and the Sustainable Development Goals	180
Table 4.2	Social impact lending CF and sustainable performance ("spillovers")	182
Table 4.3	Banks and crowdfunding	187
Table 5.1	Questions to aid in the assessment of a fund manager's ESG management system	225
Table 5.2	Investment focus of some world-famous social VC funds	237

CHAPTER 1

Socially Responsible Investment (SRI): From Niche to Mainstream

Francesco Gangi, Nicola Varrone, and Lucia Michela Daniele

1.1 Beyond the Separation Model

For decades, a fundamental question has plagued economic studies: Should economics, as a social science, address economic problems separately from ethical and moral issues? In other words, can *homo economicus*—"economic man"—an abstract unit of statistics as defined by Pareto (1951), exclude moral criteria into his decision-making? Is profit maximisation the only measure of rational conduct?

After many years of debate, and many significant events, not least the dramatic financial crisis of 2008, we can confidently answer the foregoing questions with a "no". That is, economic science, in whatever field it is applied to, from the functioning of the markets to managerial action and financial decision-making, is impoverished if it does not integrate its economically rational scientific assessments with ethical assessments aimed at giving a different meaning to the so-called economy of well-being.

As Sen (1987) asserts, a perspective that restricts itself to Pareto optimality as its sole criterion is too narrow and cannot contribute to the effective advancement of knowledge and interpretative models useful for social well-being. Such assumptions about the perfect functioning of markets along with hypotheses of absolute rationality and competitive equilibrium as necessary and sufficient conditions for the economy

F. Gangi et al., *The Evolution of Sustainable Investments and Finance*,
https://doi.org/10.1007/978-3-030-70350-9_1

of well-being have great limits which have conditioned the development of economic theories, including liberalism (Friedman 1962). These assumptions have been laid as the foundation stones of some of the most successful theoretical frameworks of the last century and have had been a major cause of the alleged separation between economic and moral action, with agency theory being a particular case in point (Friedman 1970; Jensen and Meckling 1976). We must therefore take this as our starting point in grasping the extent of the challenge inherent in recognising the role of ethics and social responsibility in the functioning of the economy, its agents and its institutions. This involves transcending the separation model identified by Freeman (1994) and reacting to the exasperated objective of separating economic actions from moral decisions.

This topic has gathered momentum in recent decades, and the contributions of new theoretical perspectives, such as stakeholder theory (Freeman 1984) and behavioural finance (Kahneman and Tversky 2000), have alerted scholars to new logics of how economic relations and financial decisions might work.

However, awareness of this issue pre-dates this by some years. Nearly seven decades ago, Pareto (1951) acknowledged the criticisms that his economic models had attracted on account of their neglect of moral and legal considerations. However, this was the result of a deliberate methodological choice: an attempt to isolate the different parts of a phenomenon in order to study them separately. What this approach bequeathed us, though, is the figure of "economic man" who acts in isolation.

Pareto (1951) recognised how the economy, once separated from ethical and legal considerations, acquires the standards proper to the formal sciences. For Pareto, economists—as scientists—must first isolate the economic phenomenon from other phenomena and then investigate them individually. This logic of separation of a complex and comprehensive phenomenon into constituent parts is a method that can be traced back to the work of Adam Smith. In fact, according to Pareto, Smith made a fundamental choice. Concerned though he was about the welfare of the community, he focused on only part of the problem and omitted such things as morality, religion and legislation, which also have important roles to play (Pareto 1951). Indeed, without adequate social organisation it is difficult to assume the well-being and wealth of a community. However, these aspects were consequently excluded from neoclassical economic thinking, and this led to another important phenomenon known as "liberal utopia", which is a product of the separation of the sub-objective of ethics. In particular, liberal utopia is based on the assumption that all practical problems in the economy can be reduced to simplified concepts such as *mine* and *yours*. The "magic formula", as Pareto defines it (1951, p. 306), for ensuring collective well-being and prosperity is the "*laissez faire, laissez passer*". This view is a partial one, and it corresponds to a methodological choice that fails to contribute to achieving the set objective. If absolute rationality and individual economic freedom are the main drivers of well-being and wealth creation for communities, then it would not explain, claims Pareto, the "large number of wretched people" that characterised the England of free trade in which economic freedom was already widespread during the advent of the capitalist model (1951, p. 307).

1.1.1 *The Separation Model and the Growth of Inequality and Imperfections*

Pareto's assertion, above, is highly pertinent for our times. An economy dedicated to free trade, and characterised by profoundly deregulated financial systems, has contributed to economic growth in many countries (Rajan and Zingales 2003), acting as a driver of new phenomena such as the globalisation of markets. Meanwhile, we have witnessed the spread of social inequality (Piketty 2020) and the growth of uncertainty linked to the bursting of speculative bubbles on financial markets. Some data may help in clarifying the extent of the contradictions in this phenomenon.

According to the NGO Oxfam (2020), the richest 20%, worldwide, own 72% of total assets, while the poorest 60% own just 12.4% of their national wealth. To offer an even clearer insight into the gap, Oxfam estimates that, in 2018, the top 26 billionaires possessed the same wealth as the poorest half of the world's population. Thomas Piketty, in his recent book on capital and ideology (2020), pointed out that the challenges of the twenty-first century are twofold: inequality and climate. Piketty notes that in less egalitarian countries (e.g. the US and the Middle East) the gap between rich and poor is almost seven times greater than in more egalitarian countries (e.g. Europe). In addition, the concentration of wealth which he calls "neoproprietarianism" (Piketty 2020, p. 742) is growing (50% of the world's poorest own just 5% of private wealth), and the gap is increasing between the average income of the richest 1% compared to 50% of the group of the poorest of world's population (the ratio is 25x in Europe, while in the US it is 80x, thus confirming the link between wealth inequality and how egalitarian the country is).

The natural environment, too, has suffered the effects of globalisation, liberalisation of markets and erosion of regulations, with a significant increase in CO_2 emissions, which have grown by 63% between 1990 and 2014, and 23% since the beginning of the new millennium (World Bank 2020). Piketty contends there is a direct relationship between lifestyle and consumption inequalities and CO_2 levels. With the US responsible for 57% of total emissions and Europe 15%, this indeed implies that climate damage is correlated with the level of inequality of wealth distribution. This inequality is ultimately associated with ethical gaps. For example, countries that have sought to abolish racial discrimination, such as South Africa or Brazil, are characterised by higher levels of inequality in wealth distribution (Piketty 2020).

In terms of institutional governance, the abrogation in 1999 of the Glass–Steagall act of 1933 has now been recognised as the precursor of an institutional crisis of ethics (Gangi 2013) which paved the way for a series of grave financial crises as a result of the bursting of the speculative Internet securities bubble (the "dot-com crash") in 2000, and the bursting of the sub-prime mortgages bubble in 2008. Along the same lines, de Bruin (2015) noted that the Depository Institutions Deregulation and Money Control Act (1980) turned the sale of sub-prime mortgages into a legitimate business. As Rajan and Zingales (2003) have admitted, deregulation has frequently been associated with bank crises and consequent financial instability. Those two authors identify a lack of competences and new knowledge as the main reason behind that particular cause-and-effect. De Bruin (2015), on the other hand, warned that a prerequisite for effective liberalisation is the assumption of greater responsibility. Hence, it can be argued that necessary conditions for economically successful deregulation and liberalisation of markets are, on the one hand, new competences and, on the other, responsibilities—at individual and institutional level. In the absence of these, deregulation and liberalisation are most likely doomed to failure, allowing the dark side of finance to rear its head (Rajan and Zingales 2003).[1] The loss of millions of jobs (unemployment in the EU grew from 7 to 11%) and the collapse of world

[1] Rajan and Zingales (2003), following the collapse of Enron, concluded that liquidity in modern financial markets, coupled with leverage and new financial instruments, serves to amplify the scope and the catastrophic effects of managerial misconduct.

GDP (e.g. −6.4% in the US; −4% in the EU) are the consequences of the 2008 financial crisis, which happened within economic models frequently based on the concept of the self-regulation of markets and the supposed absolute rationality of free economic action, with no consideration of the necessary competences and responsibilities (Fig. 1.1).

A comparison of Anglo-Saxon countries before and after the 2008 financial crises highlights the evolution of the regulation of markets. In the two main Anglo-Saxon financial systems (US, UK), we can observe a significant decline in the regulation index before the sub-prime bubble burst and a rebound during the years following the financial crises.

Assumptions about the efficiency of self-regulating markets (Jensen 1978) have been deeply embedded for a long time. Several decades ago, one of the greatest experts on economic crises, John Maynard Keynes, in one of his radio speeches, later transcribed in The Listener (1934), in finding the theory of the automatic adjustment of economic systems insufficient, admitted that he was siding with the "heretics". There are too many internal and external disturbing factors, shocks and errors,

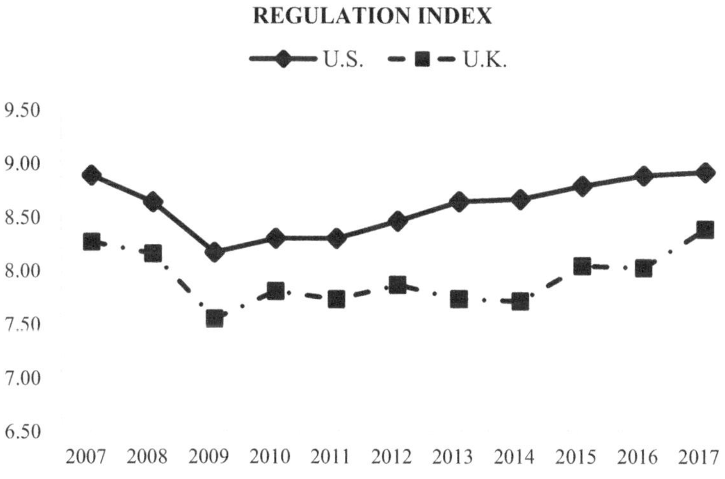

Fig. 1.1 Comparison of Regulation Index (*Source* Fraser Institute)

which includes the malfunctioning in the practice of capitalist individualism already criticised by Marx (1876) and identified by Sombart as a rational and ordering force proper to the bourgeois spirit (1974).[2]

Over time, we have seen an increase in the interference caused by errors and the creaking of an economic model based on an alleged separation between economic and non-economic considerations, such as respect for moral values. In fact, the US Financial Crisis Inquiry Commission of 2008, investigating the most serious crisis since 1929, acknowledged that the crisis was determined not only by economic factors but by a perfect storm caused by a combination of an irresponsible use of leverage, inadequate governance systems, and the systematic collapse of social responsibility and ethics (Financial Crisis Inquiry Commission 2011). At this point, it should be made clear that ethics pertains to individuals, while social responsibility concerns organisations. Indeed, following Sciarelli, ethics aims to reconcile economic objectives with corporate social responsibility and represents a unitary model of behaviour, functional to the development of the enterprise and to long-term entrepreneurial goals (Sciarelli 1999). Therefore, the separation model, as hypothesised by Freeman (1994), which was, to a certain extent, a provocative response to Goodpaster's (1991) criticism of the stakeholder paradox, contains a fundamental integral flaw which dates back to specific methodological choices in neoclassical economic studies. In other words, this model excludes ethics on both a conceptual and an operational level, and also social responsibility, with respect to the dimension of economic action and financial decisions (Harris and Freeman 2008; Freeman and Auster 2011). At the same time, this methodological individualism lays the foundations for an alleged axiological neutrality which places social relations in a market economy outside the boundaries of any judgements around ethics and economics. According to this perspective, "economic man" (e.g. manager, entrepreneur or investor) is called to act by choosing, in a rational way, the option that will likely maximise his personal interest. On the one hand, this interest is combined with profit maximisation as a

[2] It is worth remembering here that there is a common thread between Marx, Sombart and Schumpeter linking their different theses of the evolution of capitalism. For Marx (1867), the relentless individualism of the capitalist would lead to the end of capitalism itself as a result of social struggles. For Schumpeter (1934, 1947), the creative destruction brought to bear by the entrepreneurial classes represents the very engine of capitalist development. At the same time, its evolution leads to a progressive separation between property and control. This is not the end of capitalism, but a new evolutionary phase.

measure of the degree of satisfaction of personal utility; on the other hand, the satisfaction of individual interest omits the external consequences of personal choices.

1.1.2 *The Separation Model Within Firm Theory: Profit Maximisation Is not a Convincing End-Goal*

At the level of business theory, Robert Solomon (1992) identifies two basic goals: the production of goods and services at a certain level of quality and the creation of profit for shareholders. When the firm is represented as a mere legal fiction in a nexus of voluntary contracts, the end-goal of profit maximisation finds one of its truest expressions in the agency perspective (Jensen and Meckling 1976). As Friedman (1970) asserted in his provocative article, a company's only social responsibility is to make profit. The task of the manager to whom the property has delegated the governance of the company is to make as much money as possible in the interest of the shareholder, but, as Carroll (2000) recognises, this must be done in compliance with social rules. Friedman also recognises the importance of laws and rules under which economic decisions are taken by corporate executives, saying "That responsibility is to conduct the business in accordance with their desires, which generally will be to make as much money as possible while conforming to their basic rules of the society, both those embodied in law and those embodied in ethical custom" (Friedman 1970, p. 1). However, consistent with Kant's proposition (1781), respect for law is not sufficient to make a man virtuous. *Rectitudo juridica* is different from *rectitudo ethica*, which pertains to the inherent goodness of a decision or action and its moral intention. Moreover, a philanthropic initiative might be a good moral decision in the perspective of society, but should be avoided since it does not meet the aims of the shareholders. For Friedman the responsibility of corporate actions must be reduced to responsibilities of human agents, but in this way, without a corporate responsibility the social responsibility of firm no longer has reason to exist (de Bruin 2015). As such, not only is the view of Friedman and his followers too narrow, but this theory, successful as it is, is based on a logic of separation between social responsibility and economic responsibility, and from this separation are derived multiple effects and limits.

A manager who invests company resources in social responsibility initiatives will end up imposing undue taxation on the shareholders' profits. Rather, maximising the net present value of cash flows, given the cost of capital, is the main rule to be followed in a risk–return trade-off environment. The shareholder must impose controls and incur charges to steer managerial action and prevent it deviating from the above objective of wealth maximisation, although the consequences of this behaviour may well have impacts beyond the confines of the manager–shareholder relationship. However, from an evolutionary perspective, the shareholder cannot be considered simply as a monolithic entity. Its utility function can have multiple facets that transcend the logic of maximum profit (Beal et al. 2005). Furthermore, "maximum profit" is an incomplete objective on both a conceptual and an operational level. What sort of profit is to be maximised: short-, medium- or long-term? Further, under what level of risk should profit be maximised? These are questions that cannot be answered by agency theory. Rather, at an operational level, many uncertainties remain around efficiency. We might consider, for example, what the implication is of a preference for maximum profit in the short term at the expense of a larger profit in the longer term. The same accusation of incompleteness that is levelled at stakeholder theory can be made about the theory of profit maximisation and the agency model—an accusation that goes to the heart of the inherent concept. Propensity or aversion to risk is a personal characteristic: it differs among individuals. Can a manager act in pursuit of the unconditional maximisation of profit without an adequate assessment of the underlying risk and the variance in risk propensity among shareholders?

Risk plays a central role in the pursuit of "optimal" solutions. As Scharding (2015) points out, any approach to risk that is not accompanied by a clear ethical vision can be immoral insofar as it does not consider the consequences that the choices made might have on financial, economic and social systems. Consistent with de Bruin (2015), the moral intensity of an action depends on the magnitude of the consequences of that action and on the probability with which these consequences may arise: in other words, the riskiness of the decision. A good example of this is the testimony by Citigroup's top managers to the US Financial Crisis Inquiry Commission that the $40 billion exposure in securitised mortgages had barely attracted their attention, and to which they had devoted no more than 1% of their time. The underlying issue here is not that there should be a denial of economic self-interest; without such economic self-interest, Sen (1987) recognises there would be no more economic exchanges and the capitalist economies would collapse. The issue is more complex, and cannot be reduced to self-interest alone. As Sen (1987) states, every individual is a citizen of the world who cannot limit the purpose of his or her behaviour to the satisfaction of the function of personal utility as measured in terms of marginal wealth. Rather, a prior evaluation of one's actions must also be made with respect for the community of which one is a part and to the well-being of which one contributes. On an individual level, the separation model, which implicitly pervades agency theory, can be adequately repudiated by the logic of the scale of individual expectations.

As Sciarelli (1999) points out, if we look at the successful entrepreneur, his/her well-being is gauged not only by the metric of maximum profit but also by the power and prestige he/she enjoys in the community in which he/she operates. Without responsibility there is no power, and without the satisfaction of social needs there cannot be prestige. In this perspective, the theory of social success combines the achievement of profit with the well-being of the community. An individual faced with economic choices cannot have profit maximisation as his/her only distinct aim. The manager is not necessarily an opportunistic individual to be monitored, whose actions lead to additional costs to transfer on the market. For these reasons, Ghoshal (2005) defines the agency model as a form of "pretence of knowledge" (von Hayek 1989). The mainstreaming of agency theory has played a part in the transformation of capitalism and contributed to the hijacking of managerial studies (Gangi and Méric 2019).

Not all individuals are equal, and separating managerial decisions from moral values is a mistake (Harris and Freeman 2008). The agency model cultivates a pessimistic view of human nature (Ghoshal 2005), with its assumption that opportunism always prevails. Within these parameters, the manager has only one frame of reference within which to ensure that personal interest takes no precedence over the interest of the company and the shareholder: the maximisation of profit. However, as we have pointed out, this is only a claim to knowledge: the belief that setting a single rule or a single objective can guarantee that opportunism is eliminated and that the satisfaction of shareholders' interests can be effectively measured. In reality, as Mulligan (1986) points out, the manager, although an individual, does not always act alone, but interacts with different bodies, such as boards of directors and shareholders' meetings. The company cannot be represented as a battlefield for the conflicting interests of the various stakeholders. Rather, the continuity of the business and its ability to achieve a sustainable competitive advantage over the long term are functions of the specific propensity to balance the interests of stakeholders (Freeman 1984; Barney 1999; Barnett and Salomon 2012). Jensen (2001) also recognises the opportunity inherent in transcending the rhetoric about the dichotomy between the moral imperative of profit and the costs of social responsibility. He proposes a new measure of business success represented by the ability to support different social demands without contravening the economic function of the company to create value in the long term. Jensen (2001, p. 9) defines this renewed vision of value creation as "enlightened value maximization".

At the firm level, the separation model is repudiated not only by the uncertainty of the risk/time combination, but also by multiple other factors that have become increasingly evident over time. First of all, even though the debate remains open, there is evidence of a positive relationship between corporate social performance and corporate financial performance (Orlitzky et al. 2003; Gangi et al. 2020a). Not only is corporate social responsibility (CSR) an essential component of the corporate function but it can also be considered a strategic driver fostering the creation of more lasting relationships of trust with stakeholders (Barnett and Salomon, 2012), putting the company on a firmer footing. CSR can be a source of reputational capital (Gangi et al. 2020b) and social capital (Russo and Perrini 2010) that fuels relationships of trust between the various parties involved. Increased trust reinforces the barriers against opportunistic behaviour and fosters knowledge sharing. The latter can become a driver for the company's improved ability to innovate (Scuotto et al. 2019).

In addition, on the financial side, the shareholding structure has changed over the years, with an increase in the proportion of institutional investors leading to a profound reorientation of the shareholder configuration. This observation is backed up by the numbers. For example, in the North American market, assets under management reached \$88.5 trillion in 2017, from \$50.7 trillion in 2007, a growth of 75% in ten years. Among institutional investors, studies (e.g. Revelli 2017; Gangi and Varrone 2018) have confirmed an increase in those who base their investment decisions according to triple-bottom-line logic (Elkington 1994). In particular, ESG (environment, social and governance) criteria are becoming increasingly common among investors: in Europe, ESG risks are being taken into account by 55% of institutional investors in their investment choices—up from 40% in 2018 (Mercer 2019). At the global level, sustainable investment activities in the five main markets (Europe, United States, Japan, Canada, Australia/New Zealand) reached $30.7 trillion at the beginning of 2018, an increase of 34% in two years (GSIA 2018). The expansion of these investors has brought shareholders' interests into closer alignment with corporate social responsibility. The world's leading banks (119 as of August 2020) have signed up to the Equator Principles, which combine environmental ratings with financial assessments (Scholtens 2009). The Bank of Italy recognises that a company's environmental responsibility is one of the risk components that must be factored in (Bank of Italy 2020). The management of climate risk by banks is not only an outcome of regulatory pressure but has also become an imperative due to increasing commercial pressure (McKinsey 2020).

Under the scenarios described above, the separation model has no reason to exist. As stated by de Bruin (2015, p. 32), "even under the austere view of the corporation as a nexus of contracts or a generator of Pareto efficient wealth, epistemic virtues are essential".[3] The diffusion of socially responsible principles in finance confirms the need for a new approach. However, to better understand the scope of this phenomenon, it is necessary to look more closely at distinctive aspects of this type of investment analysis, by placing economic decisions, ethical principles and corporate social responsibility in the same context.

1.2 The Main Features of Ethical and Socially Responsible Investing

The rise of a new form of shareholder represented by the ethically and socially responsible investor is one of the factors contributing to the repudiation of the separation model. If, as we have documented, the proportion of this type of investor is increasing, then a manager cannot limit his/her initiatives exclusively in pursuit of profit maximisation. The creation of firm value is not sufficient in itself; rather, it is also necessary to demonstrate *how* this value is created, with regard to ethical and social values.

[3] According to the Kantian definition, ethics is the doctrine of virtues (*Lectures on Ethics*, 1781).

As we have suggested—and we will discuss this in more detail over the following pages—data from the market corroborate the thesis of the proliferation of the ethical and socially responsible investor, strongly represented among institutional investors. In Europe, for example, institutional investors account for over 90% of socially responsible investing (SRI) (Eurosif 2018). This percentage has shown a constant growth over the years, even when the ability to raise capital was negatively affected by the financial crises. Parallel with the consistent rise in the number of socially responsible investors,[4] and the increased demand for CSR from various stakeholders, the attention paid by scholars to this important phenomenon has also increased over time.

Scientific interest has focused on three main areas: the logic of capital allocation; the financial performance of socially responsible investment funds (SRFs), in comparison with conventional funds; and the determinants of the choices of individual investors.

As far as a definition is concerned, the literature frequently refers to the absence of consensus about a general definition framing this particular type of investor, and notes the contrast between the classical approach based on Markowitz's (1952) position and that of Moskowitz (1972), which highlights the positive impact on information asymmetries linked to the application of the more selective ESG criteria. Indeed, as we will show later, the distinction between SRFs and conventional funds is tending to blur: among the latter we can also find the application of risk–return maximisation, optimal diversification and ESG integration (Gangi and Varrone 2018).

At the end of the 1990s, when the phenomenon was beginning to take off, Cowton (1999) identified the socially responsible investor as one who consciously uses traditional financial criteria alongside ethical, social and environmental principles. He/she is an investor who, either by mandate or by individual choice, sees an investment decision as an act that is not morally neutral (Dembinski et al. 2003).

During the first years of the new millennium, there is a growing awareness that we are encountering investors who are applying their own personal values. Michelson et al. (2004) also identify these investors as

[4] In this chapter, consistent with Renneboog et al. (2008), the terms "ethical investors" and "socially responsible investors" are interchangeable, since they frequently overlap. The same applies for ethical and socially responsible investment funds, usually simply referred to as "SRFs".

those who consciously integrate their own personal values with social considerations and economic factors (Michelson et al. 2004). Probably one of the most comprehensive analyses is that of Cheah et al. (2011), who focus on the act of ethical and socially responsible investment and define it as the philosophy and practice of making strategic investment decisions by integrating financial and non-financial considerations, including personal values, social demand, environmental concerns and corporate governance issues.

1.2.1 Capital Allocation by Socially Responsible Funds

The literature considers SRFs to be specific "financial products" which have a mandate to pursue ethical, social, environmental and governance objectives in addition to financial returns (Cowton 2004, p. 249). This specificity is translated into capital allocation logics based on screening criteria for target companies that distinguish them from conventional funds. The most commonly found screening criteria (Renneboog et al. 2008; Cortez et al. 2009; Ferruz et al. 2012) are social, ethical, environmental and corporate governance screens. According to a broader definition of sustainable investment, further classifications can be made to identify a more precisely articulated set of investment strategies (Scholtens 2014; GSIA 2018), namely: *exclusions*, *best-in-class investment selection*, *norms-based screening*, *integrated approach*, *engagement*, *sustainability-themed investment* and *impact investment*.

The oldest is the *exclusion-based strategy* or negative screening. It consists in excluding, from the eligible investment universe, companies, sectors and countries that are involved in controversial or harmful activities (so-called sin stocks) such as abortifacients, contraceptives, fur, genetic engineering, stem cells, animal testing, weapons, pornography and tobacco. Moreover, among the increasingly prevalent types of negative screens, Eurosif (2018) and recent academic literature (e.g. Henriques and Sadorsky 2017; Trinks et al. 2018) signal the emergence of fossil fuel divestment, wherein fossil fuel holdings are replaced by climate-friendly assets (Ayling and Gunningham 2015; Scholtens 2017; Trinks et al. 2018). Emblematic in this regard is the decision in 2018 by the world's largest sovereign wealth fund, the Government Pension Fund of Norway (Statens Pensjonsfond Utland), to divest 4% of its portfolio allocated in securities of companies active in the extraction and sale of oil and gas. This is a very important issue that will characterise future economic scenarios; substantial estimates have been posited of a decrease in the value of investment in fossil fuels of even more than 40% (HSBC 2013). The new sustainable development scenario maps out a way to meet sustainable energy goals in full, requiring rapid and widespread changes across the energy system. Such a scenario pursues a path fully aligned with the Paris Agreement by holding the rise in global temperatures to "well below 2°C ... and pursuing efforts to limit [it] to 1.5°C", and meets objectives related to universal energy access and cleaner air (IEA 2019). These scenarios therefore outline the new operational boundaries within which institutional investors interested in social and environmental criteria can move.

Nevertheless, the exclusion-based approach remains the one most adopted by SRFs (Eurosif 2018), although authors have frequently raised criticism about the exclusion of controversial activities in investing processes (Trinks and Scholtens 2017). For example, Richardson (2013) contends that the application of negative screening may be not consistent with an SRF's fiduciary mandate, to the extent to which sin stocks may offer a better financial performance than socially responsible ones. De Colle and York (2009)'s argument, on the other hand, is that the exclusion of companies involved in controversial industries may not accurately reflect investors' values and beliefs. The authors suggest that, instead of *a priori* industry exclusions, a more effective approach to SRI would be to focus on the actual impacts of the company's products on society, also taking into account the relationship that the company creates with stakeholders and the context in which it operates. Other studies have highlighted the influence that the application of negative screening has on portfolio returns. In this regard, while some authors suggest that mutual funds investing in sin stocks offer positive expected returns (e.g. Hong and Kacperczyk 2009; Liston and Soydemir 2010), others indicate that the application of negative screening to stock picking does not affect financial performance (Humphrey and Tan 2014; Lobe and Walkshäusl 2016).

Alongside the exclusion approach, SRFs may adopt so-called positive screening. This strategy involves selecting sectors, companies or projects that respect minimum standards according to specific ethical and ESG dimensions. Positive screening is often applied alongside "best-in-class investment selection", which focuses only on those stocks offering the best socially responsible performance in a certain industry, sector or category. Both negative and positive screening may be driven by a "norms-based approach", wherein the mutual fund's manager selects those companies that are aligned with international standards related to socially responsible investments, such as the UN Global Compact or the OECD Guidelines.

However, according to Eurosif (2018), the "integrated approach" is the investing strategy with the highest growth rate (+27%) between 2015 and 2017. This, also termed the "triple bottom line approach" (Renneboog et al. 2008), combines rigorous economic analysis with the application of both negative and positive screening. Indeed, as reported by Peifer (2014), the majority of SRFs opt for strategies based on screening, with only a few being actively engaged in socially responsible investing.

"Engagement" is an investment strategy in which SRFs assert themselves through active ownership, voting in shareholders' meetings or undertaking activities (i.e. emails, telephone or direct conversations with senior management) to convince the board of directors to improve a firm's socially responsible performance. Johnsen (2003) and de Colle and York (2009), among others, argue in favour of SRFs' engagement in active ownership as a more effective way of inducing companies to undertake CSR initiatives. In this regard, several studies in recent years have demonstrated SRF engagement having a positive impact on the CSR outcomes of target firms (Scholtens and Dam 2007). For example, Chen and Yang (2020) show that the level of institutional ownership and shareholder attention to ESG issues can potentially affect the firm's engagement in CSR activities. Focusing on the firms listed on the Palestine Stock Exchange, Zaid et al. (2020) find that institutional ownership positively affects CSR disclosure. The authors also point out that the relationship is more positive for firms with high levels of board independence. Other studies have investigated whether the adoption of ESG engagement strategies by institutional investors enhances shareholder economic value; Kim and Lyon (2011), for example, report an increase in shareholder value for companies participating in the Carbon Disclosure Project. A few years later, analysing an extensive sample of US public companies between 1999 and 2009, Dimson et al. (2015) demonstrate that successful ESG engagements have a positive impact on abnormal returns. The authors also note that engagements are more likely to be successful with regard to governance issues than to environmental and social ones. Flammer (2015), on the other hand, focuses on the relationship between shareholders' proposals regarding CSR and corporate financial performance.

Specifically, analysing a sample of 2,729 CSR shareholder proposals from 1997 to 2012, she notes that close-call proposals lead to superior financial performance. More recently, Hoepner et al. (2018) find that successful ESG engagements can reduce firms' downside risk. They also reveal that the relationship between engagements and firms' downside risk changes across ESG themes, showing a higher impact when engagements are in relation to environmental issues.

In addition to the above-mentioned strategies around SRI, prior literature signals the rise of "sustainability-themed" investments (Ortas et al. 2014), wherein the logics underlying the capital allocation processes are driven by one or more specific sustainable themes. This approach implies that the mutual fund's manager includes in the investment portfolio those companies that operate consistently with a specific thematic strategy. According to Eurosif (2018), in recent years themes related to environmental issues (i.e. environmental protection, energy efficiency, sustainable transport, waste management) have gained momentum in mutual funds' capital allocation processes, resulting in the proliferation of green funds (Allevi et al. 2018). On this topic, Ielasi and Rossolini (2019) suggest that the growing interest of institutional investors in environmental areas may be explained by two main factors. First, civil society's growing concerns over environmental threats such as climate change, deforestation and natural resource depletion translate into an increasing demand for environmentally sustainable financial products. Second, in order to mitigate the impact of environmental risks on the portfolio's financial performance, institutional investors are increasingly oriented towards investing in environmentally sustainable assets.

Finally, there is "impact investing", which is among the strategies least applied by socially responsible investors (GSIA 2018). It is an approach that consists in allocating capital to investments that have a positive environmental and social impact, without ignoring the financial returns. Prior literature suggests that impact investing may assume a complementary role alongside philanthropy and government support (Freireich and Fulton 2009), since it has the potential to create environmental and social value.

However, regardless of the above classification, recent trends and norms show the fiduciary mandate of SRFs based on socially responsible screens to be the standard. The 2015–2019 United Nations Environment Programme Finance Initiative (UNEP FI) report is clear on this point, stating that the fulfilment of fiduciary duties between savers and institutional investors requires them to incorporate ESG issues into investment analysis and decision-making processes. Furthermore, the report makes it explicit that "investors that fail to incorporate ESG issues are failing their fiduciary duties and are increasingly likely to be subject to legal challenge" (UNEP FI 2019, p. 2). Thus, following the UNEP FI, socially responsible investment criteria are not merely an option; rather, they represent a prerequisite in achieving the financial goals of investment decisions. To address this point, the following section will look at SRI as a mainstream phenomenon.

1.2.2 Decades of Studies on SRI Performance

Over the past decades, SRI has garnered increasing interest from the academic community (Ortas et al. 2014; Widyawati 2020). In particular, ever since Moskowitz's pioneering study (1972), the conversation has largely centred on the financial performance of socially responsible investing (e.g. Pereira et al. 2019). Along this line, most of the literature has investigated if, and to what extent, the application of socially responsible screens to capital allocation processes leads to superior risk-adjusted returns compared to unscreened investments (e.g. Bauer et al. 2005; Kreander et al. 2005; Renneboog et al. 2008; Pereira et al. 2019).

From a theoretical point of view, scholars have brought forth a number of arguments to shed light on how the integration of socially responsible issues in investment processes impacts on financial outcomes. Among the most frequent theses used to support the underperformance hypothesis of SRIs, Modern Portfolio Theory (Markovitz 1952; Roy 1952) avers that the application of any screens that limit the spectrum of eligible investments will result in less efficient portfolios. More specifically, according to this theoretical perspective, an investment portfolio may be characterised by two kinds of risk: namely, systematic and idiosyncratic risks. The former refers to the volatility of the entire market and the latter to each security taken individually. Modern Portfolio Theory contends that idiosyncratic risk can be eliminated from a portfolio through the random selection of securities from the entire investment universe. This approach, known as diversification, permits the construction of optimal investment portfolios that provide the lowest degree of risk for a given level of return.

Following this line of reasoning, the application of SRI filters to the capital allocation process limits the investment universe to only those securities that present favourable socially responsible profiles, limiting diversification opportunities (e.g. Kurtz 1997). As a consequence, the capital allocation logics underlying SRIs constrain portfolio risk optimisation, implying a higher idiosyncratic risk, not compensated by higher returns. In this regard, Geczy et al. (2003) find that the cost of including SRI issues in investment processes may vary from a few to over 30 basis points, depending on the type of investors and the fund managers' skills. Moreover, Cortez et al. (2009) and Areal et al. (2013) assert that, in addition to offering more limited diversification potential, SRIs incur additional costs in monitoring the social performance of target investments, which further detracts from the returns of the investment portfolio.

An alternative stream of research argues that Modern Portfolio Theory focuses only on the reduced diversification of SRIs, and fails to take into account the different values that each security can bring to the investment portfolio (Moskowitz 1972). That is to say, according to SRI advocates, even though SRIs imply the restriction of the investment universe, SRI fund managers are focusing on better-managed firms, which tend to achieve superior financial performance (e.g. Barnett and Salomon 2006). This view sinks its roots deep into stakeholder theory (Freeman 1984), according to which those firms that properly manage their stakeholders' interests are more likely to be successful in the long run (e.g. Donaldson and Preston 1995). There is a broad recognition in prior literature that firms with superior social performance tend to establish better relationships with stakeholders, which in turn favours the achievement of competitive advantage (Porter and Kramer 2006) and better economic results (e.g. Turban and Greening 1996; Barnett and Salomon 2012). In this sense, the adoption of socially responsible initiatives represents an opportunity for companies to build trust and long-term relationships with stakeholders (e.g. Waddock and Graves 1997; Jensen 2001). Thus, engagement in socially responsible initiatives may be interpreted as a signal of high-quality management, one that knows how to improve its relationships with the firm's key stakeholders (e.g. Cortez et al. 2009). With this in mind, the application of socially responsible screens should enable investment in companies that have good relationships with their stakeholders, and are thus more likely to present superior financial performance (Bollen 2007). On the same lines, Barnett and Salomon (2006) suggest that the better financial outcomes of socially responsible firms selected by SRI fund managers translate into higher financial performance at the portfolio level.

Conversely, other authors have also argued that engagement in socially responsible initiatives is costly (e.g. McWilliams and Siegel 1997; Jensen 2002) and may penalise firms' profit, potentially putting those firms with positive social performance at a competitive disadvantage compared to the less socially responsible firms. According to this view, SRI portfolios may therefore present worse risk-adjusted results than conventional ones, as a consequence of investing in firms with positive socially responsible conduct but with poor financial performance.

With these theoretical arguments as the background, scholars have often analysed the financial performance of SRI portfolios in comparison with their conventional peers (e.g. Pereira et al. 2019). Apart from a few exceptions which find a positive relationship between social screens and a portfolio's financial performance (e.g. Humphrey and Lee 2011), the vast majority of the empirical evidence points to the neutral performance of SRI portfolios (Hamilton et al. 1993; Goldreyer and Diltz 1999; Statman 2000; Bauer et al. 2005, 2006, 2007; Bello 2005; Renneboog et al. 2008; Cortez et al. 2009).

As suggested by recent studies (e.g. Galema et al. 2008; Sandberg et al. 2009; Ferruz et al. 2012; Cortez et al. 2012), these inconclusive results may be due to the scant attention paid to the heterogeneity of the social screens. In fact, effects on financial performance may vary greatly according to the type and intensity of socially responsible screens applied in the capital allocation processes (Barnett and Salomon 2006; Humphrey and Tan 2014). Sandberg et al. (2009) suggest that SRI heterogeneity may be a reflection of strategical, practical and cultural differences. Influenced by this idea, scholars have gradually shifted their emphasis from the question "does it pay to be good?" to "when does it pay to be good?" (King and Lenox 2001; Capelle-Blancard and Monjon 2014).

Following this logic, the primary line of empirical research has focused on types of screening strategy. Humphrey and Lee (2011), for example, in their analysis of a sample of Australian SRI funds, find that negative screening significantly increases the portfolio risk and reduces opportunities for portfolio diversification, while positive screening significantly reduces the portfolio risk. More recently, Auer and Schuhmacher (2016) demonstrate that negative environmental and social screens have no significant impacts on performance, while positive screens negatively affect the portfolio value. Capelle-Blancard and Monjon (2014), on the other hand, show that positive screening has no impact on a portfolio's financial performance.

Some authors have focused on screening intensity instead. Here, Barnett and Salomon (2006) found a curvilinear (U-shaped) relationship between the number of screens applied by mutual funds and a portfolio's financial returns. Renneboog et al. (2008) indicate that a higher intensity of social and governance criteria reduces a mutual fund's returns. A few years later, Lee et al. (2010), focusing on a sample of SRI funds from the United States, find that the number of screens applied has no influence on the portfolio's unadjusted returns or idiosyncratic risk.

Drawing on the idea that specific dimensions of social responsibility may have differentiated effects on mutual fund financial performance, other studies have focused on specific ESG factors. For example, Borgers et al. (2015) investigate the influence of social screens on risk-adjusted returns, while Capelle-Blancard and Monjon (2014) demonstrate that sectoral or transversal screens can influence mutual funds' performance. A growing number of empirical studies in this stream of research have focused on green or environmental screens. Among the most recent, Silva and Cortez (2016) find that European and US green funds underperform their conventional peers during non-crisis periods and periods of low interest rates. They also find that green funds show a similar performance to funds certified with an SRI label. More recently, Ibikunle and Steffen (2017), focusing on a sample of European mutual funds over the period 1991–2014, find that the risk-adjusted returns of green funds have improved over time, outperforming conventional funds over the time window of 2012–2014.

Further, this area has seen increased attention from researchers on the financial performance of mutual funds that are divesting from fossil fuels. For example, Henriques and Sadorsky (2017) find that portfolios that disinvest from fossil fuels and utilities, investing instead in clean energy initiatives, outperform portfolios that continue to invest in fossil fuels and utilities. At the same time, Trinks et al. (2018) note that divestment from fossil fuel companies does not seem to penalise portfolio performance. Overall, these new trends confirm the escalation of factors that are driving socially responsible investment and indicate the business opportunities available in responding with financial products tailored to savers' growing demands for sustainability.

1.2.3 *The Drivers and Characteristics of the Individual Socially Responsible Investor*

With years of empirical evidence failing to reach consensus about the financial performance of SRFs—which is partly attributable to methodological issues—there is a topical interest in looking in more depth at the characteristics and motivations that drive individual investors towards socially responsible investment choices.

The literature has evolved significantly over time and has investigated various aspects in order to illuminate the behaviour of the individual investor. Survey techniques vary: from interviews to experimental approaches with the administration of investment options. An emerging feature is the heterogeneity of investors, who, it is increasingly clear, do not represent a monolithic category (Sandberg et al. 2009).

Towards the end of the last century one of the first studies to be conducted using interviews—by Mackenzie and Lewis (1999)—identified the "mental account" approach (Thaler 1985), whereby an ethical investor would invest only a small part of their investments in ethical portfolios, leaving the remainder in conventional products. A few years later, Webley et al. (2001), using an experimental approach, stressed the importance of personal ethical values in investment decisions. In particular, individuals who declare themselves to be ethical investors are, generally, genuinely committed to ethical responsible investments and tend to affirm their choice even if such investments perform poorly. The picture is varied. Dembinski et al. (2003) distinguish between two ethical approaches: value-based and impact-based. In the former, moral factors prevail, leading to the exclusion of certain investments or encouraging intermediaries and companies to adopt practices that conform with personal ethical standards (value-based ethics). In the latter, economic motivations prevail, which are satisfied by obtaining a higher financial return in the medium to long term from investments in companies that safeguard values (impact-based). This duality is consistent with Etzioni's assertion (1988) that the economy has a moral dimension that transcends the separation model. The importance of personal ethical values is confirmed by the study of McLachlan and Gardner (2004), which finds significant differences between socially responsible investors and conventional investors, according to personal belief in ethical issues and on their moral intensity. This personal-values dimension expands the investor's utility function. Beal et al. (2005) identify a holistic utility function (U) which includes the variable of the degree of ethicality of the investment (e):

$$U = f(E_R, \sigma_R, e) \qquad (1.1)$$

Where (E_R) is the expected return, while (σ_R) is the standard deviation. Assuming a condition of risk aversion, the first variable has a positive impact, while the second a negative impact on the utility function. In this case, however, the indifference map is more complex and assumes a three-dimensional configuration (indifference plane) in order to include a preference towards the ethical component of the investment in addition to the classic risk–return trade-off. In particular, the hypothetical non-maximisation of wealth of the socially responsible investment, due to its potentially suboptimal diversification, can be compensated by incremental benefits of a psychological nature that fill any gap caused by lack of expected financial return. Moreover, the objective of stimulating positive change in society through socially responsible investment translates into the possibility of enjoying a personal benefit indirectly linked to the results obtained thanks to the activities undertaken by socially responsible companies that are worthy of being selected for the investment. This mix of factors is well supported by the concept of investment as a potential vehicle for social change and it is also consistent with a view of investment motivation as a "pleasure stream" (Beal et al. 2005, p. 72) derived from a range of benefits: financial, psychic and social.[5] Obviously, the intensity of benefits linked to ethical values and social objectives varies from individual to individual, which justifies efforts to identify segmental clusters within this specific category of investors. Nilsson (2009) identified three different segments: one mainly concerned with profit, one mainly concerned with social responsibility, and one concerned with both financial returns and social responsibility. In the third case, the investor assesses both financial results and the consideration being made towards

[5]The work of Beal et al. (2005) on the psychological benefits of ethical investment is rooted in behavioural finance studies that recognize the possibility of non-rational choices by the investor. In particular, the approach that combines the classical utilitarian perspective with expressive benefits is based on the concepts of anchoring (Tversky and Kahneman 1974), framing (Slovic 1995), cognitive bias and emotion effect investor (Shefrin and Statman 1985).

social responsibility in deciding how to allocate his or her resources to socially responsible investments.

The mix of extra-financial benefits has a psychological dimension, and for this reason they are referred to by the literature as "expressive benefits" (Glac 2009). These are benefits linked to the act of investing being conceived in a socially responsible way as an extension of the investor's identity and personal values. From this perspective, in addition to the conviction that one is behaving correctly, and doing good in social and environmental terms, socially responsible investment acquires a high symbolic and experiential value (Gangi et al. 2016).

Some important considerations emerge from the foregoing excursus. First, the variety of possible personal expectations and intangible benefits related to SRI might imply that the expectations of the ethical investor can be placed along a continuum ranging from a strictly ethical perspective to a strictly financial orientation (Pasewark and Riley 2010). In fact, the behaviour of this particular category of investor is attributable to various motivations, such as financial return, non-wealth return and contribution to social change.

Second, the practical and scientific recognition of the ethical and socially responsible investor contributes to the repudiation of the separation model—the dichotomy between the pursuit of profit and respect for ethical values—thereby contributing to a wider value creation objective. In fact, even if there is segmentation within this category of investor, the intensity of the ethical component that guides investment choices shows that the behavioural model theorised by Jensen and Meckling (1994) is not always applicable, as this investor does not always appear willing to exchange his/her ethical "need" for the "want" of a better risk–return combination (Berry and Yeung 2012). Awareness of these realities has a practical implication. On the one hand, investment fund management teams are offering more socially responsible financial products. On the other hand, companies, in their attempt to attract these investors and to establish more stable relationships with their stakeholders, have intensified social disclosure, adhering to standards and protocols aimed at a greater diffusion through this new channel of communication between the company and its internal and external community.[6]

[6]For example, in July 2020, companies who had signed up to the United Nations Global Compact (UNGC) for the dissemination of social communication came from 157 countries and published a total of 72,277 public reports.

1.3 The Fiduciary Duties of SRFs: The Nexus Agency Model

In this section we need to return to agency theory (Jensen and Meckling 1976). Agency theory, as we have seen, is an interpretative model applicable to social phenomena that involve cooperation between several parties in order to achieve a pre-established goal. The relationship between the parties involves a division of tasks and roles: one party, the principal, instructs the other party, the agent, to perform a service in the interests of the former, in return for payment. The performance of the assigned task presupposes a certain degree of decision-making autonomy on the part of the agent. As we will explain later, the theoretical construct of the agency model can also be applied to the relationship between an individual saver and the management of socially responsible investment funds.

First of all, it should be pointed out that agency theory assumes that the relationship between agent and principal is adversely affected by a number of "imperfections" which can make the contract between the parties less efficient, both in terms of cost and in terms of achieving the principal's objectives. These imperfections need careful scrutiny in order to understand the determinants of a less economically efficient allocation of capital, as well as the extra-financial expectations of the socially responsible investor.

The first potential imperfection factor is a divergence of interests between the principal and the agent, both of whom are committed to the maximisation of their own utility function. A divergence of interests leads to contractual inefficiency in so far as the agent, in order to satisfy his/her own interests, takes decisions contrary to those of the principal. The manager is in a position of information advantage, which may be abused to achieve personal objectives to the detriment of the principal.

The second potential factor is the low observability of the agent's actions, which are often verifiable only *ex post*, when it may be too late for them to be corrected. The theme of observability is central to the theory of imperfect contracts, which is also known as the Grossman–Hart–Moore (GHM) model (Grossman and Hart 1982; Hart and Moore 1990). In common with agency theory, this model makes reference to contractual logic for the governance of rights and duties, and to the theme of information asymmetries. Contractual incompleteness rests on two basic principles. The first is the principle of non-verifiability. According to this principle, the parties involved in a transaction are not always able to agree *ex ante* about all possible actions to be taken and how to use resources. The second is the principle of descriptive incompleteness. That is, since contracts are written by human beings, who by definition are subject to limited rationality (Simon 1947), contractual relationships cannot contain an unequivocal and exhaustive description of all the activities that might have to be undertaken, the results that can be achieved and the modalities of their attribution between the parties. As such, the GHM model offers a further key to understanding the limits of the principal–agent relationship, by which, rooted in the principle of limited rationality, a further increase in the level of inefficiency of the agency relationship might be explained. It should be noted that the GHM model is in fact complementary to both the theory of limited rationality and to agency theory. The theory of contractual incompleteness rests on a twofold imperfection: the unavailability of information and insufficient processing capacity. The GHM model considers not only the lack of information but, importantly, the impossibility of transcribing the information that is available in an exhaustive way within complete contracts.

A third potential imperfection factor is post-contractual opportunism. That is, the agent, by exploiting the principal's inability to monitor effectively, and the resultant information asymmetry, may reduce his/her commitment to satisfying the principal's interests or may seek hidden benefits beyond those formally provided for in the contract. Such post-contractual opportunism is a form of moral hazard.

The fourth and final factor, closely linked to the first three, concerns the costs of the agent's monitoring and incentivising. In other words, given the potential divergence of interests, the non-observance of the agent's performance and the risks of moral hazard, control systems and additional benefits must be put in place for the agent. Both measures involve "agency costs", thereby making the management of the principal–agent relationship less efficient. Indeed, due to agency costs, the welfare of the principal is reduced and his/her utility function is not optimally fulfilled.

The agency model can be of use in analysing some aspects of the relationship between the management of SRFs and the savers who entrust to these funds a part of their wealth in order to pursue financial and extra-financial objectives. The savers, or unit holders, give a mandate to invest their financial resources following certain standards that should be governed by the fund's operating rules and mission. At the same time, the investment fund management gives a mandate to the managers of the target companies to invest. This dual configuration is what the literature calls a *nexus agency model* (Black 1992; Woidtke 2002) and it can amplify the imperfections of the agency relationship. In addition to the fiduciary relationship between saver and fund management, there is also a fiduciary relationship between the fund management and the management of the target company. There is a consequent risk of a duplication of the imperfections and costs to be transferred to the market.

The saver is in a position of information asymmetry with respect to the fund management team. This may encourage opportunistic behaviour if, for example, fund managers intend to pursue financial objectives at the expense of non-financial objectives (Barnett and Salomon 2006). Second, the performance of the fund management is not easily observable by the saver; savers are generally only able to verify the fund management's activities *ex post*. Moreover, given the increased complexity of the utility function of the ethical and socially responsible saver, the monitoring requirements can be more complex. The fund management will have to meet expectations not only of financial returns but also those regarding the social and environmental issues that confer the saver's expressive benefits. This implies a dual level of information and monitoring that may be too onerous or not easily achieved. On the fund management side, the role is characterised by a duality: to the saver it acts as an agent, and to the management of the target company it acts as a principal. The management of the SRF selects the companies in which to invest following ethical and social screening criteria and will be required to verify the extra-financial performance consistently against the mandate received from the saver. The fund management will operate in information asymmetry with respect to the management of the target company and this may imply additional agency costs. Therefore, the nexus agency model—between saver and SRF, and between SRF and the target companies—can multiply imperfections and agency costs to the detriment of contractual efficiency, as follows:

$$AC_{S/SRF} + AC_{SRF/Target\,Firm} = CI \tag{1.2}$$

Where $AC_{S/SRF}$ indicates the agency costs between savers and the SRF management, $AC_{SRF/Target\,Firm}$ indicates the agency costs between the SRF management and the management of the target firm, while CI indicates the contractual inefficiency of the nexus agency model.

As we will explain later, not only is the nexus agency model a potential source of contractual inefficiency, but it can contribute to an increase in the risks associated with the mainstreaming of financial products inspired by SRI (Revelli 2017).

1.4 The Opportunities and Risks of SRI as a Mainstream

As anticipated in Sects. 1.2 and 1.3 above, the investment process of SRFs can usefully be viewed through the lens of the nexus agency model (Black 1992; Woidtke 2002). This model successfully captures the twofold fiduciary mandate, which involves individual investors, SRFs and firms in the context of SRIs. The first fiduciary mandate concerns the individual investors and managers of the SRFs, wherein the former engages the latter to allocate the fund's resources in socially responsible companies. In the second fiduciary mandate, the managers of the SRFs engage the managers of the target companies to undertake CSR initiatives. Thus, according to the nexus agency model, there are two levels of fiduciary duties to be fulfilled. On the one hand, the SRFs' managers are required to include in the investment portfolio those firms presenting favourable corporate financial performance (CFP) and corporate social performance (CSP), in order to satisfy both the financial and extra-financial expectations of individual investors (e.g. Bollen 2007). On the other hand, the managers of target companies are called on to engage in socially responsible initiatives in order to meet the SRI investment screening criteria adopted by SRFs.

As recognised in the prior literature (Schneider 2002), this dual fiduciary mandate can amplify the possibility of agency problems (Ross 1973; Jensen and Meckling 1976), hindering the fulfilment of the fiduciary duties. The extra agency level may mean a duplication of information asymmetries and agency costs, increasing the exposure of the individual investor to the risk of the SRF's fiduciary mandate not being fulfilled. According to Rhodes (2010), socially responsible investing, through SRFs, represents a "multilayered information asymmetry problem". The author observes that, in order to ensure that an individual investor's resources are invested consistently with her/his financial and extra-financial expectations, two conditions are necessary. First, firms need to provide SRFs with reliable information about their engagement in socially responsible initiatives; second, SRFs need to provide individual investors with reliable information about the socially responsible performance of the firms in the investment portfolio.

In this regard, scholarly attention over the last few years has turned to the motivations that may make SRFs abandon their fiduciary duties and thus create a discrepancy between their actual objectives and the expectations of individual investors interested in socially responsible products (e.g. Kempf and Osthoff 2008). Among some of the most prominent theses on this topic we find the argument that SRFs might in fact apply investment strategies that are similar to those of conventional funds (Glassman 1999; Goetz 1997). SRFs are suspected of deliberately failing to use rigorous social and environmental screening criteria in the investment process in order to obtain financial performance more aligned with that of conventional funds (Barnett and Salomon 2006). This approach appears to be in alignment with the predominance that financial objectives tend to assume in the context of socially responsible investments, and grafts onto the debate around the transition of SRIs from niche to mainstream. As the prior literature would have it (Azoulay and Zeller 2006; Revelli and Viviani 2015), while the "SRI niche" refers to the socially responsible dimension as a relevant and constraining factor in investment management processes, the "SRI mainstream" seeks only to integrate socially responsible issues into conventional investment approaches. Following this line of thought, a transition to the mainstream risks the prioritisation of the financial stakes and a concomitant loss of SRI's true ethical identity (Revelli 2017).

From the nexus agency model perspective, the dual fiduciary mandate may contribute to increased risks in connection with the mainstreaming of SRI. The additional agency costs and information asymmetries that are inherent within the nexus agency model might influence SRFs to favour investment processes that prioritise financial objectives to the detriment of SRI's ethical dimension. This approach may be galvanised further by the limited monitoring capabilities of the individual investor, whose choice is constrained to the small number of SRFs being offered on financial markets, and who has no influence over the investment management team's selection of target companies. Individual investors' informational disadvantage is further exacerbated by a lack of transparent and reliable ESG ratings (Windolph 2011; Widyawati 2020) that might bear witness to the actual environmental and social performance of the SRFs.

Although a number of authors have pointed to the growing dissemination of SRI (Alda 2019; Revelli 2017), others question whether SRI actually has the potential to switch from niche to mainstream (Gutsche and Zwergel 2020). On one side, there are some signals that corroborate SRI's tendency to move towards a mainstream. For example, in a recent survey by the Censis Institute, 52.3% of Italians declared an interest in ESG investments (Censis 2020). Moody's observed that the growing interest in social issues and the current favourable regulatory framework may lead to an exponential growth of ESG assets (Moody's 2020). Recent academic literature (Alda 2019) demonstrates a growing adoption of ESG criteria by conventional funds. On the other side, there are authors making the point that SRI still represents a small proportion of the financial market compared to conventional investments (e.g. Gutsche and Zwergel 2020), and that there are several barriers hindering the transition to mainstream. Juravle and Lewis (2008), for example, cite individual, institutional and organisational impediments, and, more recently, Gutsche and Zwergel (2020) cite individual investors who, despite their intentions of investing in a socially responsible way, were discouraged from actually allocating resources in SRIs because of the high information costs and their distrust towards providers of financial products.

The existence of these two opposing views testifies to the fact that a consensus around the implications and real growth potential of SRI is still far from being reached (Widyawati 2020). Future research may empirically verify whether the rise of SRIs is correlated with a loss of their ethical identity (Revelli 2017). Moreover, future studies might investigate the effects that upcoming ESG regulations (e.g. the EU Taxonomy[7]; see below) are having on the dissemination of SRIs. A third area in which future research might expand our understanding is to enquire whether the disclosure of transparent information about SRFs' integration of socially responsible issues in the capital allocation process incentivises individual investors to allocate resources in SRIs.

1.5 ESG Rating of Investment Funds

The recent and rapid growth of the SRI market (Eurosif 2018; GSIA 2018) has stimulated a demand for indicators that measure socially responsible performance. Indeed, the inclusion of socially responsible issues in investment capital allocation processes represents one of the

[7] https://ec.europa.eu/info/business-economy-euro/banking-and-finance/sustainable-finance/eu-taxonomy-sustainable-activities_en.

main drivers behind the growing demand for standardised ESG information (Avetisyan and Hockerts 2016; Dilla et al. 2016). Institutional investors need objective ESG metrics in order to select socially responsible companies and to manage the assets in the portfolio. Just as financial ratios reflect financial performance, it seems to be a fundamental requirement that social responsibility, too, should be expressed in measurable indicators. Against this backdrop, ESG rating agencies have emerged in response to the demands of SRI investors for reliable indicators of firms' engagement in socially responsible activities (e.g. Escrig-Olmedo et al. 2019).

In a similar vein to credit rating agencies, which evaluate ability to fulfil financial obligations, ESG rating agencies analyse businesses and provide indicators on companies' socially responsible behaviour. Rating agencies thus reduce the information asymmetries between companies and stakeholders interested in the firm's socially responsible profile (Shafer et al. 2005; Scalet and Kelly 2010). These stakeholders, which include customers, employees and investors, might not in fact have any access to useful, genuine information on corporate engagement in socially responsible initiatives (Ramus and Montiel 2005), so ESG rating agencies emerge as third-party intermediaries who possess the resources and knowledge to assess and inform about corporate social performance. For the most part, each agency develops its own methodology and calculates ratings taking into account public information, company reports and data collected via questionnaires.

These ratings play a crucial role in capital allocation decision-making, informing investors, and facilitating comparisons among the socially responsible performance of target investments. As documented in recent literature (Escrig-Olmedo 2019; Drempetic et al. 2019), mutual funds' SRI strategies are largely based on ESG ratings. Chelli and Gendron (2013) contend that ESG agencies' rating mechanisms can alter corporate orientation towards socially responsible issues, influencing if and how capital flows to economic activities.

Recently, alongside their company assessments, some rating agencies have started providing ESG ratings at the mutual fund level (e.g. Morningstar's Sustainability Rating; MSCI ESG quality score), which has expanded the accessibility of ESG evaluations to all investors. While company ESG ratings are mostly used by institutional investors, the introduction of mutual funds' ESG scores opens the door to greater adoption by retail investors as well (Ammann et al. 2019).

As these ratings become more common in financial markets, they have attracted greater attention from researchers who have analysed their characteristics (e.g. Avetisyan and Hockerts 2016; Muñoz-Torres et al. 2019) and questioned the validity of their metrics (e.g. Drempetic et al. 2019). While acknowledging the methodological advances made by rating agencies over the last decade (Escrig-Olmedo et al. 2019), most studies find ESG ratings subject to a number of problems (e.g. Saadaoui and Soobaroyen 2018). For example, doubts have been cast about the ability of ESG rating agencies to effectively distinguish socially responsible firms (Entine 2003; Avetisyan and Hockerts 2016). For Windolph (2011) and Widyawati (2020), the main concerns about ESG metrics include: a lack of transparency, doubts about the independence of ESG agencies and the absence of consistency among ratings coming from different agencies.

With regard to lack of transparency, authors have noted that the methods and criteria used to calculate ratings mainly remain opaque or confidential (e.g. Schäfer 2005; Windolph 2011; Chelli and Gendron 2013), so it is difficult to fully understand how ESG agencies make their assessments (Scholtens 2009). Prior studies suggest that this lack of transparency may be partly attributable to the logic of competition: disclosing their methodologies exposes rating agencies to the risk of being imitated by competitors (e.g. Delmas and Blass 2010; Boiral and Heras-Saizarbitoria 2020). Scalet and Kelly (2010) observe that the negligible transparency about methods and assumptions behind the ESG ratings may be a cause of concern for stakeholders, who need reliable metrics to evaluate corporate social performance. Other authors (e.g. Avetisyan and Ferrary 2013; Avetisyan and Hockerts 2016) have noted that ESG rating agencies often provide company ratings that have been solicited and paid for by the same company, which creates the conditions for conflicts of interest, calling the independence of the raters into question.

With regard to the lack of consistency, scholars have observed that this may be attributable either to the absence of a standardised methodology (Chatterji et al. 2016; Dorfleitner et al. 2015) or else to the difficulties of aggregating very different data (Windolph 2011; Boiral and Heras-Saizarbitoria 2020): criteria and weightings underlying ESG ratings may vary significantly across agencies. Escrig-Olmedo et al. (2010) blame a lack of consensus about the definition of socially responsible performance (McWilliams and Siegel 1997; Lopez et al. 2007) for the non-convergence of ESG ratings. Saadaoui and Soobaroyen (2018), analysing several CSR rating agencies, find methodological differences regarding both the use of the environmental, social and governance themes and the weights attributed to each. Chatterji et al. (2016), too, find a lack of consistency among the ratings provided by six ESG agencies, although these authors note that the discrepancies are more likely due to problems with the validity of the assessments rather than a lack of convergence among the CSR conceptualisations being used. However, as noted by Schaltegger and Wagner (2017), this lack of consistency could also reflect an absence of standardisation around corporate social disclosure practices: since ESG ratings are partially based on companies' self-reported information, variations among companies in their accounting and reporting of social performance must surely affect the quality and the comparability of ESG agencies' ratings. In a recent paper, Berg et al. (2020) distinguish several drivers of ESG rating divergence, the main ones being, first of all, the broad range of attributes (i.e. exposure to carbon risk, labour practices, gender diversity) used to calculate ESG scores, and, second, the variation in approaches to measuring the attributes.

However, these aforementioned issues represent only part of the challenge facing ESG agencies. For example, prior literature (e.g. Escrig-Olmedo et al. 2019) also questions whether and, if so, to what extent, a single indicator can accurately measure a complex (Moon et al. 2005) and dynamic phenomenon (Carroll 1999) such as corporate social responsibility. Some studies observe that single scores can place too much weight on the environmental, social or governance pillar, thereby failing to capture the integrative and multifaceted nature of ESG performance (e.g. Schäfer et al. 2006). Furthermore, single ESG scores may paint an inaccurate picture of CSR, with poor performance in one area of corporate social responsibility seemingly glossed over by good performance in another.

All these issues may result in metrics that mislead their intended users. In fact, the low reliability of ratings calls into question their usefulness, compelling us to ask why stakeholders interested in socially responsible issues persist in using them. Delmas and Blass (2010) note that SRI investors may experience a loss of confidence in applying SRI screens based on unreliable ESG metrics. Avetisyan and Hockerts (2016) suggest that these issues around ESG ratings signal a need for further research in this area. Nevertheless, the ratings are widely used in research and assume a crucial role in orienting capital allocation processes (Escrig-Olmedo et al. 2019), and yet only a few papers have analysed the reliability of different competing indicators and their methodological approaches (e.g. Sharfman 1996). Further research in this area could help to clarify how accurately ESG ratings measure engagement in socially responsible activities.

1.6 Institutional Recognition of SRI: EU Regulatory Norms

The significant and steady growth of SRI (Eurosif 2018; GSIA 2018) signals, on the one hand, society's growing interest in socially responsible issues, and, on the other hand, the need for legislation regulating this financial market segment. The academic literature emphasises the necessity of harmonised regulatory frameworks to move sustainability from rhetoric to action (Sjåfjell and Wiesbrock 2015; Ahlström 2019). Some studies (e.g. Yamahaki and Frynas 2016) suggest that regulation on ESG will encourage institutional investors to integrate socially responsible issues in their investment processes. According to Moody's (2020), consumer preferences and regulations are the two main drivers for the enduring growth of ESG investments. Conversely, regulatory uncertainty may contribute to the increased exposure of investment portfolios to carbon-intensive assets, threatening hopes of a transition to a sustainable society (Monasterolo et al. 2019). As noted by Sherwood and Pollard (2018), the lack of legislative provisions incentivises companies to withhold ESG-related information, making it difficult for investors to identify truly sustainable activities.

However, recent years have seen a growing number of initiatives by governments and institutions around the world aimed at promoting the transition to a more sustainable society. In 2015, 193 governments signed the UN Global Agenda 2030, which defines 17 Sustainable Development Goals (SDGs) to be achieved by 2030. The Agenda covers issues ranging from ending poverty and hunger to reducing inequality and ensuring clean water and affordable clean energy. In 2016, the Paris Agreement set environmental targets to reduce levels of greenhouse gas emissions and limit global warming to 2 °C or below relative to pre-industrial levels.

Regarding the financial sector, according to MSCI the number of ESG regulations in the world increased from 64 in 2017 to 175 in 2018. At a global level, especially noteworthy is the partnership between the United Nations Environment Programme and the financial sector (UNEP FI), which aims to encourage the adoption of sustainability principles by financial institutions. Initiatives supported or co-created by the UNEP

FI include the Principles for Responsible Investment (2006), the Principles for Sustainable Insurance (2012), the Principles for Positive Impact Finance (2017) and the Sustainable Stock Exchanges Initiative (2012).

In the European context, the European Green Deal[8] outlines the EU's strategy to tackle sustainability challenges. According to the European Commission, the Green Deal "aims to transform the European Union into a fair and prosperous society, with a modern, resource-efficient and competitive economy".

In 2017, as part of the Green Deal roadmap, and in order to regulate sustainable finance in Europe, the European Commission appointed a High-Level Expert Group (HLEG) on Sustainable Finance. The Group has two main objectives: (i) to improve the contribution of finance to sustainable and inclusive growth; and (ii) to incorporate ESG issues into investment decision processes. Consistent with the European Commission's mandate, in January 2018 the HLEG published its final report, *Financing a Sustainable European Economy*, outlining the formulation of a European strategy integrating sustainability issues into the financial sector. The report's recommendations with regard to SRI include, first, the adoption of a "Sustainability Taxonomy" at the European level, which establishes which activities, investments and assets are "sustainable". Second, the Group explicitly clarifies institutional investors' duties around the integration of ESG issues within capital allocation processes. Additional regulations are required, it notes, since current financial services directives (i.e. MIFID II, Solvency II, IORP II) pay little attention to such duties. Third, the Group calls attention to the requirement for new reforms around sustainability disclosure by firms and financial institutions, in order to improve the comparability of non-financial reporting across economic sectors and financial products. Fourth, in order to increase the flow of capital in socially responsible financial products, the Group calls for increased efforts by the European Securities and Markets Authority (ESMA) and financial institutions to inform retail investors about the sustainable profile of financial products, the objective being, by providing a range of simple and understandable metrics, to enhance these investors' awareness of the sustainability impact of their investments and saving choices. Academic literature (Monasterolo and Raberto 2018) recognises that an improvement in the quality of disclosure about

[8] https://ec.europa.eu/info/strategy/priorities-2019-2024/european-green-deal_en.

environmental issues can be a main driver in encouraging investment in socially responsible assets.

In March 2018, drawing from the recommendations provided by the HLEG, the European Commission adopted an Action Plan for sustainable finance, which set out the European strategy for sustainable finance and defined ten actions to be implemented by the end of 2019. The Commission appointed a Technical Expert Group (TEG) for assistance on some of the plan's key points. Among these, one of the most important and most urgent is the establishment of an EU classification system for sustainable activities (Action Plan: Financing Sustainable Growth 2018, p. 4).

In line with the HLEG's first recommendations, and following the indications reported in the Action Plan, the TEG, supported by the ESMA and by the European Insurance and Occupational Pensions Authority (EIOPA), published a report in March 2020 which provides the methodological framework for the implementation of the EU sustainable finance Taxonomy. Because of the complexity involved in developing a taxonomy that embraces both environmental and social issues, the EU Commission decided, as a first step, to focus on only six environmental objectives, namely: (a) climate change mitigation; (b) climate change adaptation; (c) sustainability and protection of water and marine resources; (d) transition to a circular economy; (e) pollution prevention and control; and (f) protection and restoration of biodiversity and ecosystems. The Taxonomy includes "performance thresholds", which aim to identify environmentally responsible activities. In order to be considered "environmentally friendly", an activity must: (i) make a substantive contribution to one of the six above-mentioned environmental objectives; (ii) do no significant harm to the other five objectives; and (iii) meet minimum international standards guidelines (e.g. OECD Guidelines on Multinational Enterprises and the UN Guiding Principles on Business and Human Rights). As reported by the TEG, the Taxonomy's thresholds and metrics, in allowing companies and investors to identify which activities, investments and assets are environmentally friendly, will help to move resources from high- to low-carbon sectors.

Although the establishment of the Taxonomy has been postponed until the end of 2020 for climate change mitigation and climate change adaptation, and until the end of 2021 for the other four objectives, related regulations have already been published in the *Official Journal of the European Union* (*OJ*). Particularly noteworthy are Regulation 2019/2088, which establishes new ESG disclosure duties for financial market participants, and Regulation 2019/2089, which defines uniform climate transition benchmarks in the EU.

These recent and rapid developments in sustainable finance reforms have stimulated the interest of academics and researchers, who have analysed the introduction of green policies and how that is affecting: firms' investments (e.g. Monasterolo and Raberto 2018); banks' lending strategies (Campiglio 2016); and firms' energy efficiency (Raberto et al. 2019).

However, as the literature suggests (Ahlström 2019; Siri and Zhu 2019), analysis of the new sustainable finance legislation remains limited. To address this gap, future research may investigate, for example, whether the new regulatory measures actually reward more sustainable companies, and to what extent they encourage higher engagement by firms in socially responsible issues. It is also crucial to understand how the new measures affect mutual funds' integration of ESG criteria into the investment process. Other studies might analyse whether compliance with the new corporate social disclosure standards favours the selection of socially responsible firms by SRFs. Finally, there might usefully be an investigation into how the identification of environmentally responsible activities using the EU Taxonomy affects ESG agencies' ratings in terms of transparency, consistency and overall quality.

1.7 SRI Market Trends: An EU and US Comparison

As we've repeatedly made clear, the SRI market worldwide is constantly expanding. The 2008 crisis was a factor in an expansion that has seen the sector constantly increase in volume over the last ten years with investors now paying increasingly greater attention to ESG risk factors. SRI products are tailored to meet the demand for more transparent and sustainable financial options. Globally, sustainable investing assets stood at $30.7 trillion at the start of 2018, a 34% increase in two years (GSIA 2018). However, the global SRI market is highly heterogeneous in terms of investment tools and strategies. According to the Global Sustainable Investment 2016–2018 report, Europe and the US represent the main regions in terms of total assets managed in the SRI segment. The importance of this phenomenon means it warrants further investigation, so there now follows a comparative analysis of the main investment strategies adopted in these two markets: European (EU) and American (US).

In the Eurozone, the SRI market offering doubled from 2010 to 2018: an irrefutable signal that this niche is no transitory phenomenon, but rather a sophisticated and multifaceted market increasingly embracing new investing trends in accordance with the growing enthusiasm of both and institutional investors and retailers for sustainable-oriented investments. In 2014, AUM volume in the Eurozone reached 11 billion, a trend that stabilised and continued with its growth over subsequent years until 2018. From 2010 to 2018, the EU SRI market grew by an average of 19%; in 2018, Eurozone AUM contained 46% of total SRI global assets (CAGR, 2016–2018: 8%) (Fig. 1.2).

In terms of investor profiles, the EU SRI market is mainly driven by institutional investors, who in 2018 represented 70% of total AUM (Eurosif 2018). However, from 2013 to 2018, the retail segment increased in weight by 3% to reach 31% of market share, in almost all the main European markets. In line with the latest Eurosif (2018) report, this indicator is particularly relevant, as the growing impact of retail investors is believed to be a critical factor in driving SRI from niche to mainstream. In terms of asset allocation, the trend remained almost unchanged from previous years, with bonds and equity mostly dividing the lion's share of the market equally, at 40 and 47%, respectively. Within bond assets, the leading categories are sovereign and corporate bonds (33 and 60%). This evidence is in line with past trends, showing a gradual reduction in equities towards an increase in bond allocation, reflecting the more prominent

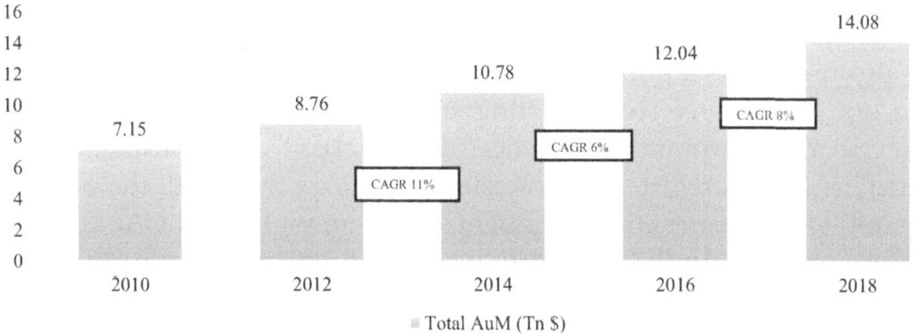

Fig. 1.2 EU SRI total market volume, 2010–2018 (*Source* Authors' elaboration of GSIA [Global Sustainable Investment Alliance] Global Sustainable Investment reports. Reports consulted: 2012, 2014, 2016 and 2018)

role given to green bonds by financial institutions such as banks (GSIA 2018) (Table 1.1).

The European SRI investment framework is extremely diversified internally.[9] In its biannual review of the market, Eurosif uses an SRI classification based on seven distinct sustainable investment strategies, namely: Sustainability-themed; Best-in-class; Norms-based screening; Exclusions; ESG integration; Engagement and voting; and Impact investing (Table 1.2).

Using total volume of AUM per year as a metric, these various SRI strategies have been observed within the 2010–2018 time frame, in order to capture their evolution in recent years in terms of market share. "Exclusion" is the principal and the oldest SRI strategy, as confirmed by its constant evolution during the time frame in question, with an eight-year compound annual growth rate (CAGR) of 23%. Despite a slight decrease between 2012 and 2014, this segment remained particularly relevant within the EU SRI market, with a market share of 41% in 2017, at that time representing the most widely adopted strategy in the Eurozone for detecting SRI opportunities. The "Engagement and voting" strategy has been particularly relevant starting from 2010 and, despite a constant slight decline, by 2018 this type of SRI growth was still in third place with respect to market share (19%). Accordingly, this trend has revived the relevance of stakeholders as a driver of SRI, with their role of influencing management. "ESG integration" registered relevant growth within the time frame being considered, signalling that the integration of sustainability criteria within financial analysis is increasingly becoming the norm as confirmed by its 2018 market share of 22%. "Norms-based screening" also represents a popular SRI strategy; this category is confirmed as being one of the preferred investing strategies in the EU, with a 2018 market share of around 14%. The "Best-in-class" and "Sustainability-themed" strategies register the lowest market share during all the periods observed; all the same, these segments show an eight-year CAGR averaging around 20%. In the Sustainability-themed category, the investments are long-term-focused and mainly overlap with the themes of the SDGs;

[9] Eurosif (2018, p. 12) defines SRI as "a long-term oriented investment approach which integrates ESG factors in the research, analysis and selection process of securities within an investment portfolio. It combines fundamental analysis and engagement with an evaluation of ESG factors in order to better capture long term returns for investors, and to benefit society by influencing the behaviour of companies".

Table 1.1 SRI investment classification: EU and US[a]

Strategy		*Description*
EU	Sustainability-themed	Sustainability-themed investments cover a wide range of themes from climate change and energy efficiency to forests and water
	Best-in-class	Best-in-class assets typically involve selecting the top percentage of companies within a sector using ESG criteria
	Norms-based screening	Norms-based screening is a strategy that involves assessing each company held in the investment portfolio against specific standards of ESG performance
	Exclusion	Exclusion or negative screening is a strategy that involves removing companies or sectors from the portfolio's investible universe
	ESG integration	ESG integration is defined as the explicit inclusion by asset managers of ESG risks and opportunities into traditional financial analysis and investment decisions based on a systematic process and appropriate research sources
	Engagement and voting	Active participation in corporate decisions and strategies
	Impact investing	Impact investing themes can broadly be classified into two categories: • Social integration, e.g. about access to affordable housing, health, finance, education, personal care or employability, to name a few examples. Microfinance would fall under this category • Sustainability-related projects in the areas of production and access to, for instance: renewable energy, food, water, sustainable agriculture. This category is heavily focused on developing markets
US	ESG incorporation only	In ESG incorporation, investment institutions complement traditional, quantitative techniques of analysing financial risk and return with qualitative and quantitative analyses of ESG policies, performance, practices and impacts
	Shareholder resolutions only	Shareholder resolutions are a meaningful way for shareholders to encourage corporate responsibility and discourage company practices that are unsustainable or unethical

[a]SRI investment strategies and related definitions were gathered, respectively, from available Eurosif and US SIF reports from 2010 to 2018

Table 1.2 Total EU SRI investment strategies by year

Year	*2010*	*2012*	*2014*	*2016*	*2018*	*Eight-year CAGR (%)*
Total AUM volume (€ trillion)[a]	*7.38*	*11.66*	*16.56*	*22.77*	*21.99*	
Time frame: 2010–2018	*Mean*		*Median*		*St. Dev.*	*14.63*
	16.11		*16.58*		*6.66*	
By strategy						
Sustainability-themed	0.09		0.06		0.06	24.76
Best-in-class	0.37		0.35		0.18	20.36
Norms-based screening	3.04		3.14		1.51	15.55
Exclusion	5.75		4.01		3.63	22.75
ESG integration	3.20		3.28		1.40	14.29
Engagement and voting	3.63		3.20		1.09	5.27
Impact investing	0.04		0.02		0.05	48.64

[a]Total AUM (trillion) reported in the this table was elaborated from Eurosif reports, 2010–2018. The totals differ slightly from the GSIA reports on account of two factors. First, Eurosif reports use data gathered from questionnaires submitted to asset managers, thus the market coverage might be not comprehensive. Second, the total volumes of AUM reported might differ as data for some European countries were not available along the entire time frame in question. *Source* Authors' elaboration of data gathered from Eurosif biennial SRI market reports. Reports consulted: 2012, 2014, 2016 and 2018

in 2018, the preferred themes related mostly to waste management (17%) and climate change (18%). Finally, the "Impact investing" SRI has only recently emerged, with a promising six-year CAGR of 48% (Fig. 1.3).

In line with the broad geography of the Eurozone, development levels and the degree of dissemination of SRI is not homogeneous throughout: differences between EU countries can be observed. France and UK taken together represented nearly 50% of total SRI assets under management in 2018, followed by The Netherlands (13%), Switzerland (12%) and Sweden (10%). Less penetration of SRI instruments is evident in the other EU countries, such as Spain, Denmark and Belgium. In some EU countries (e.g. France, UK, The Netherlands), therefore, the SRI market seems to have reached a higher degree of maturity than others (e.g. Spain, Germany, Denmark), where SRI products still play a marginal role (Fig. 1.4).

By contrasting SRI strategies in terms of volume of assets and countries' SRI market profile, it is possible to decipher how the different

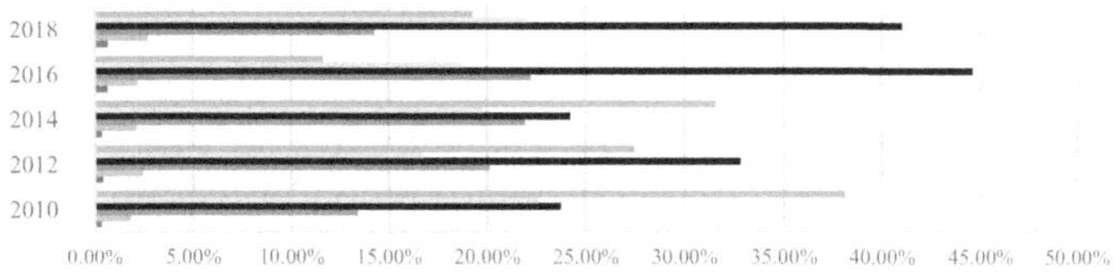

	2010	2012	2014	2016	2018
Engagement and Voting	38.11%	27.48%	31.59%	11.62%	19.27%
ESG integration	22.62%	16.73%	19.78%	18.75%	22.09%
Exclusion	23.72%	32.84%	24.20%	44.58%	41.01%
Norms-based screening	13.41%	20.12%	21.94%	22.24%	14.29%
Best-in-class	1.80%	2.43%	2.13%	2.17%	2.66%
Sustainbility themed	0.34%	0.41%	0.36%	0.64%	0.68%

Fig. 1.3 Total EU SRI investment strategies by yearly market share (*Source* Authors' elaboration of data gathered from Eurosif biennial SRI market reports. Reports consulted: 2012, 2014, 2016 and 2018)

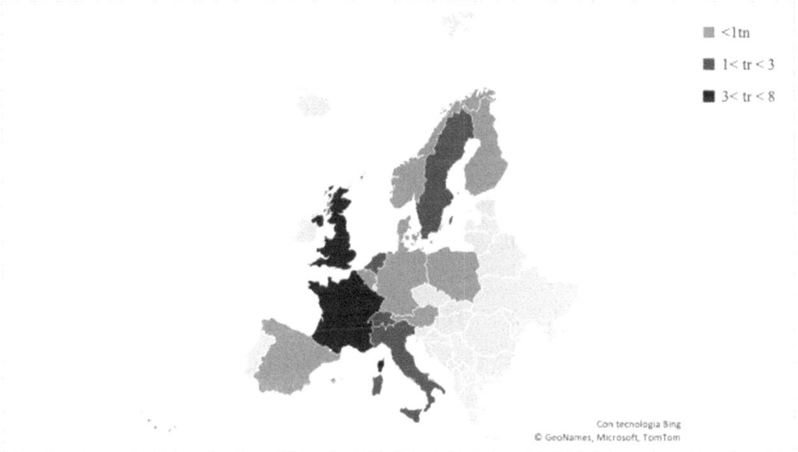

Fig. 1.4 EU SRI: diffusion in the Eurozone, 2018 (*Source* Authors' elaboration of GSIA report 2018. European SRI Study 2018)

investment approaches evolved in each European country from 2010 to 2018. The "Best-in-class" strategy has a strong presence in three of the 14 EU countries studied, with higher rates of consolidation in France, Germany and The Netherlands. "Sustainability-themed" is the leading

SRI strategy in The Netherlands, while the "Norms-based" strategy is strongest in France. "Exclusion" is popular chiefly in four countries: Italy, The Netherlands, Switzerland and the UK. The "Engagement and voting" strategy has a consolidated position in mature markets, such as the UK and The Netherlands. "ESG integration" is most prevalent in the Eurozone's main markets: namely, the UK and France. Finally, and interestingly, the "Impact investing" strategy is still gathering momentum, with an overall impact from 2012 to 2018 in a total of four EU countries: Denmark, Italy, The Netherlands and the UK (Table 1.3).

Turning to the US SRI market, data gathered from the GSIA 2018 report indicate that sustainable investing in the US is continuing its expansion at a healthy pace. In this region, the total AUM relying on sustainable strategies grew from nearly 9 trillion in 2016 to 12 trillion at the beginning of 2018: an increase of 37%. US market trends show SRI gaining momentum starting from 2014, a trend that stabilises somewhat over subsequent years with a growth rate of around 33–37% (figures for 2016 and 2018). In 2018, 39% of total global SRI assets could be found in AUM in the US market (CAGR, 2016–2018: 17%) (Fig. 1.5).

As far as investment strategies are concerned, US SIF reports use a different SRI classification from Eurosif, distinguishing simply between "ESG incorporation only" and "Shareholder resolutions only". Looking at the time frame in question, 2010–2018, in the US SRI market the majority of investments fall into the "ESG incorporation" category, which in 2018 represented 87% of total share of that market (Table 1.4).

1.7.1 The US SRI Market

In terms of themes and topics within the "ESG incorporation" category, the US SIF reports distinguish between ESG criteria adopted by money managers and those of institutional investors. An investor can incorporate ESG criteria into an investment portfolio using ESG integration, but there are different ways of doing this. Looking at ESG dimensions in terms of asset and market share (US SRI market, 2016–2018), it can be observed that assets under both money managers and institutional investors were increasingly being aligned with the ESG dimensions of product development. In that period, SRI investments in sustainable products increased by 14%, with institutional investors showing the greater interest. Breaking down the factors inherent in "ESG", while money managers tend to invest equally in social and governance issues, institutional investors consider

Table 1.3 EU SRI: total AUM by region and strategy, 2010–2018

	Market share in the EU countries by investing strategy								
EU countries by SRI investing strategy, 2010–2018	*Median total AUM (€ trillion)*	*Sustainability-themed/thematic funds (%)*	*Best-in-class (%)*	*Norms-based screening (%)*	*Exclusion*	*Engagement and voting (%)*	*ESG integration (%)*	*Impact investing (%)*	*Eight-year CAGR by EU country[a] (%)*
Austria	0.065	0.4	0.8	0.2	0.5	0.1	0.0	0.1	47.79
Belgium	0.392	2.4	1.7	1.0	3.3	0.1	1.9	0.7	8.02
Denmark	0.417	1.3	0.9	4.4	3.1	3.1	1.1	15.0	−3.96
Finland	0.227	0.3	1.2	2.0	1.2	1.1	0.7	0.2	n.a.
France	3.875	16.9	23.7	41.6	6.7	0.8	39.3	1.9	9.42
Germany	0.321	6.8	20.0	0.6	8.9	0.0	0.6	5.4	31.33
Italy	1.100	13.6	1.8	8.8	11.4	1.6	1.7	27.2	25.41
Netherlands	2.801	20.5	39.1	17.2	13.7	18.5	13.9	24.3	12.64
Norway	0.980	0.6	1.2	11.3	6.0	3.7	0.8	0.0	n.a.
Poland	0.002	0.9	0.1	0.1	0.0	0.0	0.0	0.0	45.75
Spain	0.125	3.1	0.4	0.2	1.7	0.3	0.5	4.5	30.97
Sweden	1.575	1.6	4.5	10.4	8.4	12.4	6.0	3.7	16.86
Switzerland	0.476	16.8	3.5	1.0	19.2	0.1	1.1	6.7	69.73
UK	3.452	15.0	1.0	1.2	15.7	58.2	32.3	10.1	21.43
Total	16.581	100.0	100.0	100.0	100.0	100.0	100.0	100.0	14.69

Time frame: 2010–2018; total AUM volume (€ trillion). [a]The eight-year CAGR was reported only for those EU countries with data available from 2010 to 2018. *Source* Authors' elaboration of data gathered from Eurosif biannual SRI market reports. Reports consulted: 2010, 2012, 2014, 2016 and 2018

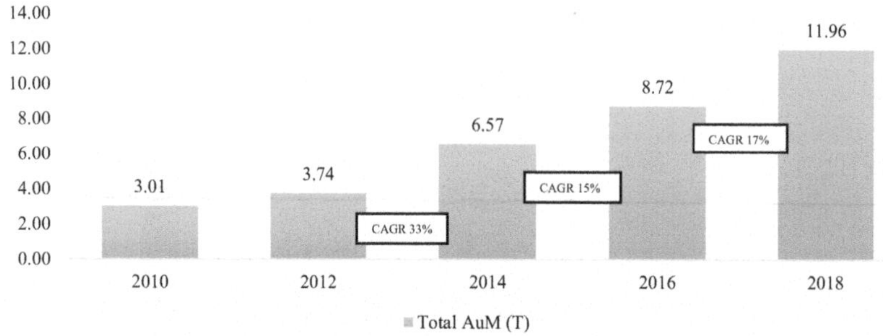

Fig. 1.5 US SRI, total market volume, 2010–2018 (biennial CAGR) (*Source* Authors' elaboration of GSIA [Global Sustainable Investment Alliance] Global Sustainable Investment reports. Reports consulted: 2010–2012, 2012–2014, 2014–2016 and 2016–2018)

Table 1.4 US SRI: investment strategy by market share, 2010–2018

Time frame: 2010–2018	*Mean*	*Median*	*St. Dev.*	*Eight-year CAGR (%)*
Total AUM volume ($ trillion)	*6.8*	*6.6*	*3.7*	*19*
By strategy				
ESG incorporation only	6.4	6.2	3.7	21
Shareholder resolutions only	1.8	1.7	0.4	2

Source Authors' elaboration of Forum for Sustainable and Responsible Investment biennial reports. Reports consulted: 2010, 2012, 2016 and 2018

the environment and governance dimensions equally important (23%) (Figs. 1.6 and 1.7).

Both the EU and the US market registered continuous growth in terms of AUM during the observed time frame. In terms of volume, while in the Eurozone sustainable investing assets registered a steady growth with an eight-year CAGR of 9%, in the US SRI growth was faster starting from 2014. At the beginning of 2018, the US SRI market started to catch up with that of the EU, as demonstrated by the differences in volume between the two regions which shrank to just a few trillion, significantly less than 2016 (Fig. 1.8).

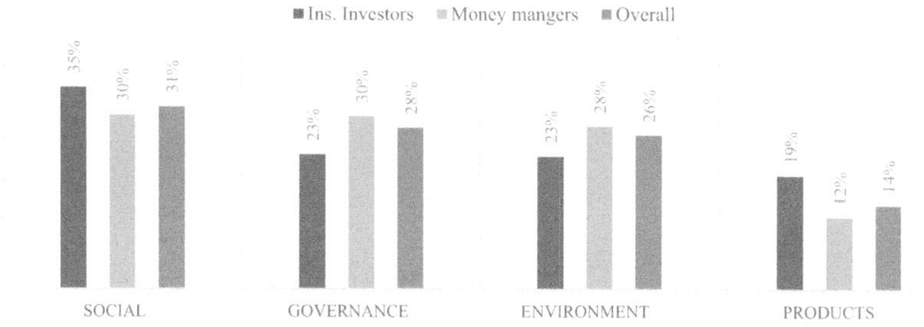

Fig. 1.6 US SRI ESG strategies: growth rate, 2016–2018 (*Source* authors' elaboration of Forum for Sustainable and Responsible Investment biennial reports. Reports consulted: 2016 and 2018)

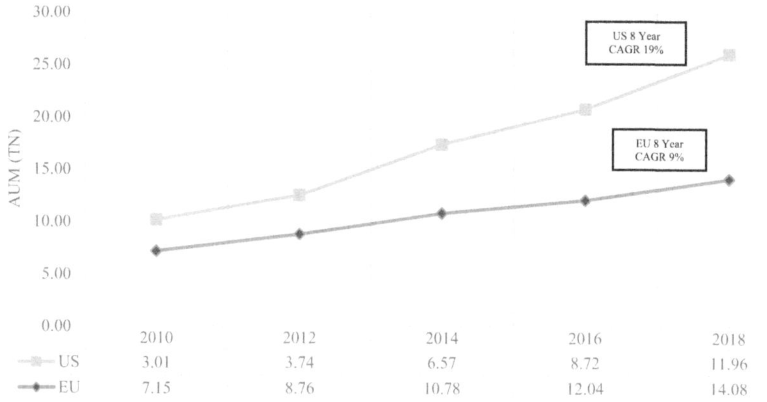

Fig. 1.7 EU and US SRI market comparison, 2010–2018 (*Source* authors' elaboration of GSIA [Global Sustainable Investment Alliance] Global Sustainable Investment reports. Reports consulted: 2012, 2014, 2016 and 2018)

Finally, with regard to the US SIF classification,[10] the "ESG incorporation only" strategy is in fact a wide umbrella that covers most of the EU SRI strategies, albeit sometimes with slightly different nomenclature: positive/best-in-class screening, negative/exclusionary screening, ESG integration, impact investing and sustainability-themed investing. In

[10] https://www.ussif.org/esg.

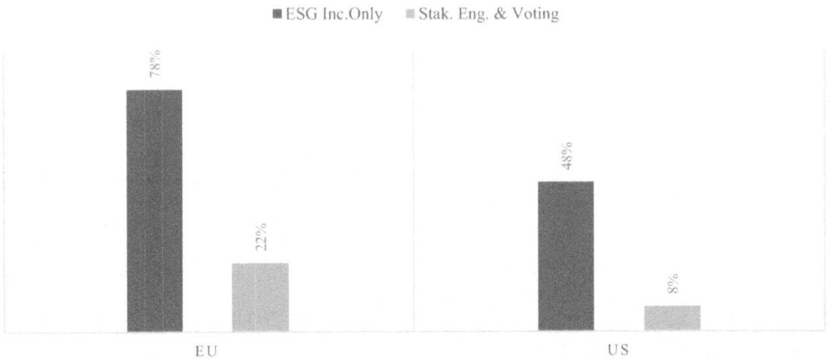

Fig. 1.8 EU and US comparison: average growth rate by investment strategy, 2010–2018 (*Source* Authors' elaboration of USSIF [The Forum for Sustainable and Responsible Investment] and GSIA reports. Reports USSIF and GSIA consulted: 2012, 2014, 2016 and 2018)

the US market from 2010 to 2018, ESG integration registered a growth of 48% on average, with the stakeholder engagement strategy developing at a slower rate (8%). This difference is more marked in the EU market, the signals from which indicate that ESG integration is widely adopted (78%) in contrast to Engagement and voting (22%).

References

Ahlström H (2019) Policy hotspots for sustainability: changes in the EU regulation of sustainable business and finance. Sustainability 11:499. https://doi.org/10.3390/su11020499

Alda M (2019) Corporate sustainability and institutional shareholders: the pressure of social responsible pension funds on environmental firm practices. Bus Strategy Environ 28(6):1060–1071. https://doi.org/10.1002/bse.2301

Allevi E, Basso A, Bonenti F, Oggioni G, Riccardi R (2018) Measuring the environmental performance of green SRI funds: a DEA approach. Energy Econ https://doi.org/10.1016/j.eneco.2018.07.023

Ammann M, Bauer C, Fischer S, Müller P (2019) The impact of the morningstar sustainability rating on mutual fund flows. Eur Financ Manage 25(3):520–553. https://doi.org/10.1111/eufm.12181

Areal FJ, Riesgo L, Rodríguez-Cerezo E (2013) Economic and agronomic impact of commercialized GM crops: a meta-analysis. J Agric Sci 151:7–33. https://doi.org/10.1017/s0021859612000111

Auer BR, Schuhmacher F (2016) Do socially (ir)responsible investments pay? New evidence from international ESG data. Q Rev Econ Financ 59:51–62. https://doi.org/10.1016/j.qref.2015.07.002

Avetisyan E, Ferrary M (2013) Dynamics of stakeholders' implications in the institutionalization of the CSR field in France and in the United States. J Bus Ethics 115:115–133. https://doi.org/10.1007/s10551-012-1386 3

Avetisyan E, Hockerts K (2016) The consolidation of the ESG rating industry as an enactment of institutional retrogression. Bus Strategy Environ 26(3):316–330. https://doi.org/10.1002/bse.1919

Ayling J, Gunningham N (2015) Non-state governance and climate policy: the fossil fuel divestment movement. Clim Policy 17(2):131–149. https://doi.org/10.1080/14693062.2015.1094729

Azoulay O, Zeller V (2006) ISR: stratégie de «niche» ou «mainstream»? Revue d'économie financière 85:191–208

Bank of Italy (2020) Annual report 2020

Barnett ML, Salomon RM (2006) Beyond dichotomy: the curvilinear relationship between social responsibility and financial performance. Strateg Manage J 27 (11):1101–1122. https://doi.org/10.1002/smj.557

Barnett ML, Salomon RM (2012) Does it pay to be really good? Addressing the shape of the relationship between social and financial performance. Strateg Manag J 33(11):1304–1320. https://doi.org/10.1002/smj.1980

Barney JB (1999) How a firm's capabilities affect boundary decisions. MIT Sloan Manage Rev 403:137

Bauer R, Koedijk K, Otten R (2005) International evidence on ethical mutual fund performance and investment style. J Bank Financ 29(7):1751–1767. https://doi.org/10.1016/j.jbankfin.2004.06.035

Bauer R, Otten R, Rad AT (2006) Ethical investing in Australia: is there a financial penalty? Pacific Basin Finane J 14(1):33–48. https://doi.org/10.1016/j.pacfin.2004.12.004

Bauer R, Derwall J, Otten R (2007) The ethical mutual fund performance debate: new evidence from Canada. J Bus Ethics 70(2):111–124. https://doi.org/10.1007/s10551-006-9099-0

Beal D, Goyen M, Philips P (2005) Why do we invest ethically? J Invest 14(3):66–77. https://doi.org/10.3905/joi.2005.580551

Bello ZY (2005) Socially responsible investing and portfolio diversification. J Financ Res 28(1):41–57. https://doi.org/10.1111/j.1475-6803.2005.00113.x

Berg F, Koelbel J, Rigobon R (2020) Aggregate confusion: the divergence of ESG ratings. MIT Sloan. Working Paper 5822-19. https://doi.org/10.2139/ssrn.3438533

Berry RH, Yeung F (2012) Are investors willing to sacrifice cash for morality? J Bus Ethics 117:477–492. https://doi.org/10.1007/s10551-012-1529-6

Black BS (1992) Agents watching agents: the promise of institutional investor voice. UCLA Law Rev 39:811–895

Boiral O, Heras-Saizarbitoria I (2020) Sustainability reporting assurance: creating stakeholder accountability through hyperreality? J Clean Prod 243:118596. https://doi.org/10.1016/j.jclepro.2019.118596

Bollen NPB (2007) Mutual fund attributes and investor behaviour. J Financ Quant Anal 42(3):683–708

Borgers A, Derwall J, Koedijk K, ter Horst J (2015) Do social factors influence investment behavior and performance? Evidence from mutual fund holdings. J Bank Financ 60:112–126. https://doi.org/10.1016/j.jbankfin.2015.07.001

Campiglio E (2016) Beyond carbon pricing: the role of banking and monetary policy in financing the transition to a low-carbon economy. Ecol Econ 121:220–230. https://doi.org/10.1016/j.ecolecon.2015.03.020

Capelle-Blancard G, Monjon S (2014) The performance of socially responsible funds: does the screening process matter? Eur Financ Manage 20(3):494–520. https://doi.org/10.1111/j.1468-036x.2012.00643.x

Carroll AB (1999) Corporate social responsibility: evolution of a definitional construct. Bus Soc 38(3):268–295. https://doi.org/10.1177/000765039903800303

Carroll AB (2000) Ethical challenges for business in the new millennium: corporate social responsibility and models of management morality. Bus Ethics Q 10(1):33–42. https://doi.org/10.2307/3857692

Censis (2020) Annual report 2020

Chatterji AK, Durand R, Levine DI, Touboul S (2016) Do ratings of firms converge? Implications for managers, investors and strategy researchers. Strateg Manage J 37(8):1597–1614. https://doi.org/10.1002/smj.2407

Cheah ET, Lindblom D, Johnson JE, Sung MC (2011) Drivers of corporate social responsibility attitudes: the demography of socially responsible investors. Br J Manage 22(2):305–323. https://doi.org/10.1111/j.1467-8551.2011.00744.x

Chelli M, Gendron Y (2013) Sustainability ratings and the disciplinary power of the ideology of numbers. J Bus Ethics 112:187–203. https://doi.org/10.1007/s10551-012-1252-3

Chen HY, Yang SS (2020) Do investors exaggerate corporate ESG information? Evidence of the ESG momentum effect in the Taiwanese market. Pacific Basin Finan J 63:101407. https://doi.org/10.1016/j.pacfin.2020.101407

Cortez MC, Silva F, Areal N (2009) The performance of European socially responsible funds. J Bus Ethics 87 (4):573–588. https://doi.org/10.1007/s10551-008-9959-x

Cortez MC, Silva F, Areal N (2012) Socially responsible investing in the global market: the performance of US and European funds. Int J Finan Econ 17(3):254–271. https://doi.org/10.1002/ijfe.454

Cowton C (1999) Playing by the rules: ethical criteria at an ethical investment fund. Bus Ethics 8(1):60–69. https://doi.org/10.1111/1467-8608.00126

Cowton C (2004) Managing financial performance at an ethical investment fund. Acc Audit Account J 17(2):249–275. https://doi.org/10.1108/09513570410532447

de Bruin B (2015) Ethics and the global financial crisis: why incompetence is worse than greed. Cambridge University Press, Cambridge

De Colle S, York JG (2009) Why wine is not glue? The unresolved problem of negative screening in socially responsible investing. J Bus Ethics 85:83–95. https://doi.org/10.1007/s10551-008-9949-z

Delmas M, Blass VD (2010) Measuring corporate environmental performance: the trade-offs of sustainability ratings. Bus Strategy Environ 19(4):245–260. https://doi.org/10.1002/bse.676

Dembinski PH, Bonvin JM, Dommen E, Monnet FM (2003) The ethical foundations of responsible investment. J Bus Ethics 48(2):203–213. https://doi.org/10.1023/b:busi.0000004598.89426.d8

Dilla W, Janvrin D, Perkins J, Raschke R (2016) Investor views, investment screen use, and socially responsible investment behaviour. Sustain Account Manage Policy J 7(2):246–267. https://doi.org/10.1108/sampj-07-2015-0066

Dimson E, Karakas O, Li X (2015) Active ownership. Rev Financ Stud 28 (12):3225–3268. https://doi.org/10.1093/rfs/hhv044

Donaldson T, Preston LE (1995) The stakeholder theory of the corporation: concepts, evidence, and implications. Acad Manage Rev 20(1):65–91. https://doi.org/10.5465/amr.1995.9503271992

Dorfleitner G, Halbritter G, Nguyen M (2015) Measuring the level and risk of corporate responsibility—an empirical comparison of different ESG rating approaches. J Asset Manage 16:450–466. https://doi.org/10.1057/jam.2015.31

Drempetic S, Klein C, Zwergel B (2019) The influence of firm size on the ESG score: corporate sustainability ratings under review. J Bus Ethics. https://doi.org/10.1007/s10551-019-04164-1

Elkington J (1994) Towards the sustainable corporation: win-win-win business strategies for sustainable development. Calif Manage Rev 36(2):90–100. https://doi.org/10.2307/41165746

Entine J (2003) The myth of social investing: a critique of its practice and consequences for corporate social performance research. Organ Environ 16(3):352–368. https://doi.org/10.1177/1086026603256283

Escrig-Olmedo E, Fernández-Izquierdo MÁ, Ferrero-Ferrero I, Rivera-Lirio JM, Muñoz-Torres MJ (2019) Rating the raters: evaluating how ESG rating agencies integrate sustainability principles. Sustainability 11:915. https://doi.org/10.3390/su11030915

Escrig-Olmedo E, Munoz-Torres MJ, Fernandez-Izquierdo MA (2010) Socially responsible investing: sustainability indices, ESG rating and information provider agencies. Int J Sustain Econ 2(4):442–461. https://doi.org/10.1504/ijse.2010.03549

Etzioni A (1988) The moral dimension: toward a new economics. Free Press, New York, NY

EUROPEAN COMMISSION (2018) Action plan: financing sustainable growth. Brussels

Eurosif (2018) European SRI study

Ferruz L, Munoz F, Vargas M (2012) Managerial abilities: evidence from religious mutual fund managers. J Bus Ethics 105:503–517. https://doi.org/10.1007/s10551-011-0982-y

Financial Crisis Inquiry Commission (2011) Financial crisis inquiry commission report—2011

Flammer C (2015) Does corporate social responsibility lead to superior financial performance? A regression discontinuity approach. Manage Sci 61(11):2549–2824. https://doi.org/10.1287/mnsc.2014.2038

Freeman RE (1984) Strategic management: a stakeholder approach. Pitman, Boston, MA

Freeman RE (1994) The politics of stakeholder theory: some future directions. Bus Ethics Q, 409–421. https://doi.org/10.2307/3857340

Freeman RE, Auster ER (2011) Values, authenticity and responsible leadership. J Bus Ethics 98(1):15–23. https://doi.org/10.1007/978-94-007-3995-6_3

Freireich J, Fulton K (2009) Investing for social and environmental impact: a design for catalyzing an emerging industry. Monitor Group, New York

Friedman M (1962) Capitalism and freedom. University of Chicago Press, Chicago

Friedman M (1970) The social responsibility of business is to make profits. New York Time Magazine

Galema R, Plantinga A, Sholtens B (2008) The stocks at stake: return and risk in socially responsible investment. J Bank Finan 32(12):2646–2654. https://doi.org/10.1016/j.jbankfin.2008.06.002

Gangi F (2013) La finanza etica come risposta alle crisi finanziarie del nuovo millennio, Guida

Gangi F, Varrone N (2018) Screening activities by socially responsible funds: a matter of agency? J Clean Prod 197:842–855. https://doi.org/10.1016/j.jclepro.2018.06.228

Gangi F, Méric J (2019) A historical perspective on the hijacking of business responsibility towards society. Business for society

Gangi F, Camminatiello I, Varrone N (2016) Analysis of private socially responsible investment: the impact of personal concern with corporate social responsibility. Rev Econ Finan 6:47–62

Gangi F, Daniele LM, Varrone N (2020) How do corporate environmental policy and corporate reputation affect risk-adjusted financial performance? Bus Strategy Environ 29(5):1975–1991. https://doi.org/10.1002/bse.2482
Gangi F, Meles A, Monferrà S, Mustilli M (2020) Does corporate social responsibility help the survivorship of SMEs and large firms? Glob Finan J 43:100402. https://doi.org/10.1016/j.gfj.2018.01.006
Geczy CC, Stambaugh RF, Levin D (2003) Investing in socially responsible mutual funds. Working Paper, Wharton School. https://doi.org/10.2139/ssrn.416380
Ghoshal S (2005) Bad management theories are destroying good management practices. Acad Manag Learn Educ 4(1):75–91. https://doi.org/10.5465/amle.2005.16132558
Glac K (2009) Understanding socially responsible investing: the effect of decision frames and trade-off options. J Bus Ethics 87:41–55. https://doi.org/10.1007/s10551-008-9800-6
Glassman JK (1999) Letting your conscience be your investment guide. Houston Chronicle, p 8
Global Sustainable Investment Alliance (GSIA) (2018) Global sustainable investment reviews 2018
Goetz T (1997) Dealing with the devil. Village Voice 19:43–44
Goldreyer EF, Diltz JD (1999) The performance of socially responsible mutual funds: incorporating sociopolitical information in portfolio selection. Manage Finan 25(1):23–36. https://doi.org/10.1108/03074359910765830
Goodpaster KE (1991) Business ethics and stakeholder analysis. Bus Ethics Q 1(1):53–73. https://doi.org/10.2307/3857592
Grossman SJ, Hart OD (1982) Corporate financial structure and managerial incentives. The economics of information and uncertainty, pp 107–140
Gutsche G, Zwergel B (2020) Investment barriers and labeling schemes for socially responsible investments. Schmalenbach Bus Rev 72:111–157. https://doi.org/10.1007/s41464-020-00085-z
Hamilton S, Jo H, Statman M (1993) Doing well while doing good? The investment performance of socially responsible mutual funds. Financ Anal J 49(6):62–66. https://doi.org/10.2469/faj.v49.n6.62
Harris JD, Freeman RE (2008) The impossibility of the separation thesis. Bus Ethics Q 18:541–552
Hart O, Moore J (1990) Property rights and the nature of the firm. J Polit Econ XCVIII:1119–1158. https://doi.org/10.1086/261729
Henriques I, Sadorsky P (2017) Investor implications of divesting from fossil fuels. Glob Finan J. https://doi.org/10.1016/j.gfj.2017.10.004
Hoepner AG, Oikonomou I, Sautner Z, Starks LT, Zhou X (2018) ESG shareholder engagement and downside risk. Working paper. https://doi.org/10.2139/ssrn.2874252

Hong H, Kacperczyk M (2009) The price of sin: the effects of social norms on market. J Financ Econ 93(1):15–36. https://doi.org/10.1016/j.jfineco.2008.09.001

HSBC Holdings plc (2013) Annual report and accounts 2013

Humphrey JE, Lee DD (2011) Australian socially responsible funds: performance, risk and screening intensity. J Bus Ethics 102:519–535. https://doi.org/10.1007/s10551-011-0836-7

Humphrey JE, Tan DT (2014) Does it really hurt to be responsible? J Bus Ethics 122:375–386. https://doi.org/10.1007/s10551-013-1741-z

IEA (2019) Global Energy & CO_2 Status Report 2019

Ibikunle G, Steffen T (2017). European green mutual fund performance: a comparative analysis with their conventional and black peers. J Bus Ethics 145:337–355. https://doi.org/10.1007/s10551-015-2850-7

Ielasi F, Rossolini M (2019) Responsible or thematic? The true nature of sustainability-themed mutual funds. Sustainability 11(12):3304. https://doi.org/10.3390/su11123304

Jensen M (1978) Some anomalous evidence regarding market efficiency. J Financ Econ 6:95–101. https://doi.org/10.2139/ssrn.244159

Jensen M (2001) In Andriof J (Ed) Value maximization, stakeholder theory, and the corporate objective function unfolding stakeholder thinking. Greenleaf Publishing

Jensen M (2002) Value maximization, stakeholder theory, and the corporate objective function. Bus Ethics Q 235–256. https://doi.org/10.2307/3857812

Jensen M, Meckling W (1976) Theory of the firm: managerial behavior, agency costs, and ownership structure. J Financ Econ 3:305–383. https://doi.org/10.1016/0304-405x(76)90026-x

Jensen M, Meckling WH (1994) The nature of man. J Appl Corp Finan 7(2):4–19. https://doi.org/10.1111/j.1745-6622.1994.tb00401.x

Johnsen DB (2003) Socially responsible investing: a critical appraisal. J Bus Ethics 43:219–222. https://doi.org/10.1023/a:1022998232503

Juravle C, Lewis A (2008) Identifying impediments to SRI in Europe: a review of the practitioner and academic literature. Bus Ethics Eur Rev 17(3):285–310. https://doi.org/10.1111/j.1467-8608.2008.00536.x

Kahneman D, Tversky A (2000) Choices, values, and frames. Russell Sage Foundation and Cambridge University Press, New York and Cambridge, UK

Kant E (1781) The critique of pure reason. Translated by JMD Meiklejohn, 2011. CreateSpace Independent Publishing Platform

Kempf A, Osthoff P (2008) SRI funds: nomen est omen. Eur Financ Manage 35(9–10):1276–1294. https://doi.org/10.1111/j.1468-5957.2008.02107.x

Kim EH, Lyon TP (2011) Strategic environmental disclosure: evidence from the DOE's voluntary greenhouse gas registry. J Environ Econ Manage 61(3):311–326. https://doi.org/10.1016/j.jeem.2010.11.001

King AA, Lenox MJ (2001) Does it really pay to be green? An empirical study of firm environmental and financial performance: an empirical study of firm environmental and financial performance. J Ind Ecol 5(1):105–116

Kreander N, Gray RH, Power DM, Sinclair CD (2005) Evaluating the performance of ethical and non-ethical funds: a matched pair analysis. J Bus Finan Account 32(7):1465–1493. https://doi.org/10.1111/j.0306-686x.2005.00636.x

Kurtz L (1997) No effect, or no net effect? Studies on socially responsible investing. J Invest 6(4):37–49. https://doi.org/10.3905/joi.1997.37

Lee DD, Humphrey JE, Benson KL, Ahn JYK (2010) Socially responsible investment fund performance: the impact of screening intensity. Account Finan 50(2):351–370. https://doi.org/10.1111/j.1467-629x.2009.00336.x

Liston DP, Soydemir G (2010) Faith-based and sin portfolios. Manage Finan 36(10):876–885. https://doi.org/10.1108/03074351011070242

Lobe S, Walkshäusl C (2016) Vice versus virtue investing around the world. Rev Manage Sci 10:303–344. https://doi.org/10.1007/s11846-014-0147-3

Lopez MV, Garcia A, Rodriguez L (2007) Sustainable development and corporate performance: a study based on the Dow Jones sustainability index. J Bus Ethics 75(3):285–300. https://doi.org/10.1007/s10551-006-9253-8

Mackenzie C, Lewis A (1999) Morals and markets: the case of ethical investing. Bus Ethics Q 9(3):439–452. https://doi.org/10.2307/3857511

Markovitz H (1952) Portfolio selection. J Finance 7(1):77–91

Marx K (1867) Capital, Volume I. International Publishers, New York

Marx K (1876) 1976 capital: Volume I

McKinsey (2020) Banking imperatives for managing climate risk. 11 June 2020

McLachlan J, Gardner J (2004) A comparison of socially responsible and conventional investors. J Bus Ethics 52:11–25. https://doi.org/10.1023/b:busi.0000033104.28219.92

McWilliams A, Siegel D (1997) Event studies in management research: theoretical and empirical issues. Acad Manage J 40(3):626–657. https://doi.org/10.5465/257056

Mercer 2019 European asset allocation survey 2019

Michelson G, Wailes N, van der Laan S, Frost G (2004) Ethical investment processes and outcomes. J Bus Ethics 52(1):1–10. https://doi.org/10.1023/b:busi.0000033103.12560.be

Monasterolo I, Raberto M (2018) The EIRIN flow-of-funds behavioural model of green fiscal policies and green sovereign bonds. Ecol Econ 144:228–243. https://doi.org/10.1016/j.ecolecon.2017.07.029

Monasterolo I, Roventini A, Foxonc TJ (2019) Uncertainty of climate policies and implications for economics and finance: an evolutionary economics approach. Ecol Econ 163:177–182. https://doi.org/10.1016/j.ecolecon.2019.05.012

Moody's—beyond passive, ESG investing is the next growth frontier for asset managers, February 2020

Moon J, Crane A, Matten D (2005) Can corporations be citizens? Corporate citizenship as a metaphor for business participation in society. Bus Ethics Q 15(3):429–453

Moskowitz M (1972) Choosing socially responsible stocks. Bus Soc Rev 1:71–75

Mulligan HF (1986) Continuing validity of state takeover statutes—a limited third generation. Notre Dame L Rev 62:412

Muñoz-Torres MJ, Fernández-Izquierdo MÁ., Rivera-Lirio JM, Escrig-Olmedo E (2019) Can environmental, social, and governance rating agencies favor business models that promote a more sustainable development? Corp Soc Responsib Environ Manag 26(2):439–452. https://doi.org/10.1002/csr.1695

NGO Oxfam (2020) Fighting inequality in the time of COVID-19. The Commitment to Reducing Inequality Index 2020. www.oxfam.italia.org

Nilsson J (2009) Segmenting socially responsible mutual fund investors: the influence of financial return and social responsibility. Int J Bank Mark 27(1):5–31. https://doi.org/10.1108/02652320910928218

Orlitzky M, Schmidt FL, Rynes SL (2003) Corporate social and financial performance: a meta analysis. Organ Stud 24(3):403–441. https://doi.org/10.1177/0170840603024003910

Ortas E, Moneva JM, Álvarez I (2014) Sustainable supply chain and company performance: a global examination. Supply Chain Manage 19(3):332–350. https://doi.org/10.1108/scm-12-2013-0444

Pareto V (1951) I sistemi socialisti. Torino, Utet

Pasewark W, Riley M (2010) It's a matter of principle: the role of personal values in investment decisions. J Bus Ethics 93(2):237–253. https://doi.org/10.1007/s10551-009-0218-6

Peifer JL (2014) Fund loyalty among socially responsible investors: the importance of the economic and ethical domains. J Bus Ethics 121:635–649. https://doi.org/10.1007/s10551-013-1746-7

Pereira P, Cortez MC, Silva F (2019) Socially responsible investing and the performance of Eurozone corporate bond portfolios. Corp Soc Responsib Environ Manage 26(6):1407–1422. https://doi.org/10.1002/csr.1756

Piketty T (2020) Capital and ideology. Harvard University Press

Porter ME, Kramer MR (2006) Strategy and society: the link between competitive advantage and corporate social responsibility. Harv Bus Rev 84(12):78–92

Raberto M, Ozel B, Ponta L, Teglio A, Cincotti S (2019) From financial instability to green finance: the role of banking and credit market regulation in the Eurace model. J Evol Econ 29(1):429–465. https://doi.org/10.1007/s00191-018-0568-2

Rajan RG, Zingales L (2003) The great reversals: the politics of financial development in the twentieth century. J Financ Econ 69(1):5–50. https://doi.org/10.1016/s0304-405x(03)00125-9

Ramus CA, Montiel I (2005) When are corporate environmental policies a form of greenwashing? Bus Soc 44(4):377–414. https://doi.org/10.1177/0007650305278120

Renneboog L, Ter Horst J, Zhang C (2008) Socially responsible investments: Institutional aspects, performance, and investor behaviour. J Bank Financ 32(9):1723–1742. https://doi.org/10.1016/j.jbankfin.2007.12.039

Revelli C (2017) Socially responsible investing (SRI): from mainstream to margin? Res Int Bus Finan 39:711–717. https://doi.org/10.1016/j.ribaf.2015.11.003

Revelli C, Viviani JL (2015) Financial performance of socially responsible investing (SRI): what have we learned? A meta-analysis. Bus Ethics 24 (2):158–185. https://doi.org/10.1111/beer.12076

Rhodes MJ (2010) Information asymmetry and socially responsible investment. J Bus Ethics 95(1):145–150. https://doi.org/10.1007/s10551-009-0343-2

Richardson BJ (2013) Socially responsible investing for sustainability: overcoming its incomplete and conflicting rationales. Transnatl Environ Law 2(2):311–338

Ross SA (1973) The economic theory of agency: the principal's problem. Am Econ Rev 63(2):134–139

Roy AD (1952) Safety first and the holding of assets. Econometrica 20(3):431–449. https://doi.org/10.2307/1907413

Russo A, Perrini F (2010) Investigating stakeholder theory and social capital: CSR in large firms and SMEs. J Bus Ethics 91(2):207–221. https://doi.org/10.1007/s10551-009-0079-z

Saadaoui K, Soobaroyen T (2018) An analysis of the methodologies adopted by CSR rating agencies. Sustain Account Manage Policy J 9(1):43–62. https://doi.org/10.1108/sampj-06-2016-0031

Sandberg J, Juravle C, Hedesström TD, Hamilton I (2009) The heterogeneity of socially responsible investment. J Bus Ethics 87:519–533. https://doi.org/10.1007/s10551-008-9956-0

Scalet S, Kelly TF (2010) CSR rating agencies: what is their global impact? J Bus Ethics 94:69–88. https://doi.org/10.1007/s10551-009-0250-6

Schäfer H (2005) International corporate social responsibility rating systems: conceptual outline and empirical results. J Corp Citizsh 20:107–120

Schäfer H, Beer J, Zenker J, Fernandes P (2006) Who is who in corporate social responsibility rating? A survey of internationally established rating systems that measure corporate responsibility. Bertelsmann Foundation

Schaltegger S, Wagner M (2017) Managing the business case for sustainability: the integration of social, environmental and economic performance. Greenleaf, Sheffield

Scharding TK (2015) Imprudence and immortality: a Kantian approach to the ethics of financial risk. Bus Ethics Q 25(2):243–265. https://doi.org/10.1017/beq.2015.17

Scholtens B (2009) Corporate social responsibility in the international banking industry. J Bus Ethics 86(2):159–175. https://doi.org/10.1007/s10551-008-9841-x

Scholtens B (2014) Indicators of responsible investing. Ecol Indic 36:382–385. https://doi.org/10.1016/j.ecolind.2013.08.012

Scholtens B (2017) Why finance should care about ecology. Trends Ecol Evol 32(7):500–505. https://doi.org/10.1016/j.tree.2017.03.013

Scholtens B, Dam L (2007) Banking on the equator. Are banks that adopted the equator principles different from non-adopters? World Dev 35(8):1307–1328. https://doi.org/10.1016/j.worlddev.2006.10.013

Schneider B (2002) Theorizing structure and agency in workplace writing: an ethnomethodological approach. J Bus Tech Commun 16 (2):170–195. https://doi.org/10.1177/1050651902016002002

Schumpeter JA (1934) The theory of economic development. Tr. By Redvers Opie. Harvard Econ Stud 40:20

Schumpeter JA (1947) The creative response in economic history. J Econ Hist 7(2):149–159

Sciarelli S (1999) Corporate ethics and the entrepreneurial theory of "social success". Bus Ethics Q 9(4):639–649. https://doi.org/10.2307/3857940

Scuotto V, Del Giudice M, Garcia-Perez A, Orlando B, Ciampi F (2019) A spill over effect of entrepreneurial orientation on technological innovativeness: an outlook of universities and research based spin offs. J Technol Transf, 1–21. https://doi.org/10.1007/s10961-019-09760-x

Sen A (1987) On ethics in economics. Basil Blackwell, New York

Shafer SM, Smith HJ, Linder JC (2005) The power of business models. Bus Horiz 48(3):199–207. https://doi.org/10.1016/j.bushor.2004.10.014

Sharfman M (1996) The construct validity of the Kinder, Lydenberg & Domini social performance ratings data. J Bus Ethics 15:287–296. https://doi.org/10.1007/bf00382954

Shefrin H, Statman M (1985) The disposition to sell winners too early and ride losers too long: theory and evidence. J Finan 40(3):777–790. https://doi.org/10.1111/j.1540-6261.1985.tb05002.x

Sherwood MW, Pollard J (2018) Responsible investing: an introduction to environmental, social, and governance investments. Routledge
Silva F, Cortez MC (2016) The performance of US and European green funds in different market conditions. J Clean Prod 135:558–566. https://doi.org/10.1016/j.jclepro.2016.06.112
Simon HA (1947) Administrative behaviour: a study of decision-making process in administrative organization. Free Press, New York
Siri M, Zhu S (2019) Will the EU commission successfully integrate sustainability risks and factors in the investor protection regime? A research agenda. Sustainability 11:6292. https://doi.org/10.3390/su11226292
Sjåfjell B, Wiesbrock A (2015) Sustainable public procurement under EU law: new perspectives on the state s stakeholder. Cambridge University Press.
Slovic P (1995) The construction of preference. Am Psychol 50(5):364–371. https://doi.org/10.1037/0003-066x.50.5.364
Solomon RC (1992) Ethics and excellence: cooperation and integrity in business. Oxford University Press, Oxford
Sombart W (1974) American capitalism's economic rewards. Failure of a dream? essays in the history of American socialism, pp 593–608
Statman M (2000) Socially responsible mutual funds (corrected). Financ Anal J 56(3):30–39. https://doi.org/10.2469/faj.v56.n3.2358
Thaler R (1985) Mental accounting and consumer choice. Mark Sci 4(3):177–266. https://doi.org/10.1287/mksc.4.3.199
The Listener (1934) Poverty in plenty: is the economic system self-adjusting? CW XIII: 485–492
Trinks PJ, Scholtens B (2017) The opportunity cost of negative screening in socially responsible investing. J Bus Ethics 140(2):193–208. https://doi.org/10.1007/s10551-015-2684-3
Trinks A, Scholtens B, Mulder M, Dam L (2018) Fossil fuel disinvestment and portfolio performance. Ecol Econ 146:740–748. https://doi.org/10.1016/j.ecolecon.2017.11.036
Turban DB, Greening DW (1996) Corporate social performance and organizational attractiveness to prospective employees. Acad Manage Rev 40:658–672. https://doi.org/10.5465/257057
Tversky A, Kahneman D (1974) Judgment under uncertainty: heuristics and biases. science 185(4157):1124–1131. https://doi.org/10.1126/science.185.4157.1124
UNEP FI (2019) Annual overview
Von Hayek FA (1989) The pretence of knowledge. Am Econ Rev 79(6):3–7
Waddock SA, Graves SB (1997) The corporate social performance-financial performance link. Strateg Manag J 18(4):303–319. https://doi.org/10.1002/(sici)1097-0266(199704)18:4%3c303::aid-smj869%3e3.0.co;2-g

Webley P, Lewis A, Mackenzie C (2001) Commitment among ethical investors: an experimental approach. J Econ Psychol 22(1):27–42. https://doi.org/10.1016/s0167-4870(00)00035-0

Widyawati L (2020) A systematic literature review of socially responsible investment and environmental social governance metrics. Bus Strategy Environ 29(2):619–637. https://doi.org/10.1002/bse.2393

Windolph SE (2011) Assessing corporate sustainability through ratings: challenges and their causes. J Environ Sustain 1(1):61–81. https://doi.org/10.14448/jes.01.0005

Woidtke T (2002) Agents watching agents? Evidence from pension fund ownership and firm value. J Financ Econ 63(1):99–131. https://doi.org/10.1016/s0304-405x(01)00091-5

WORLD BANK GROUP (2020) Data World Bank, 2020

Yamahaki C, Frynas JG (2016) Institutional determinants of private shareholder engagement in Brazil and South Africa: the role of regulation. Corp Gov 24(5):509–527. https://doi.org/10.1111/corg.12166

Zaid MAA, Abuhijleh STF, Pucheta-Martínez MC (2020) Ownership structure, stakeholder engagement, and corporate social responsibility policies: the moderating effect of board independence. Corp Soc Responsib Environ Manag 27(3):1344–1360. https://doi.org/10.1002/csr.1888

CHAPTER 2

ESG Criteria in Alternative Investments

Dario Salerno

2.1 Introduction

Traditionally, environmental, social and governance (ESG) investments have been oriented towards liquid assets such as securities and bonds. This is for a variety reasons which include the disclosure of information, the weight of assets under management, and rights provided to shareholders. Nowadays, the importance of ensuring that ESG criteria are properly incorporated into alternative asset classes has increased significantly. This is mainly because the integration of environmental, social and good corporate governance aspects into investment decisions is associated with risk reduction and value creation. Consideration of risk management as a whole and opportunities to seize market and growth opportunities is an ineluctable part of any analysis of ESG factors.

In this chapter, we explore the nexus between ESG criteria and alternative investments and shed some light on the past, present and future evolution of this phenomenon. In the first section, we provide an overview of alternative investments in which we distinguish between the different categories and highlight the role that they play in the strategic investment process. In the second section, we discuss the relationship between ESG factors and alternative investments. The latter provide companies not only with financial resources but also managerial and technical skills. In particular, the strategic interventions that

F. Gangi et al., *The Evolution of Sustainable Investments and Finance*,
https://doi.org/10.1007/978-3-030-70350-9_2

come with alternative investments aim to increase the quality of a portfolio company's management by making direct changes therein. As such, alternative investments are closely related to ESG criteria, which interrogate issues related to environmental sustainability, community relations, and best corporate governance practices. In the third section, we discuss ESG investment policy in the private equity (PE) industry. In contrast with other alternative investment categories, the PE industry was demonstrating high growth in ESG investing even before the financial crisis. PE investors routinely use at least some ESG criteria as filters in the screening process. In the fourth section, we will focus on the ESG due diligence process—an in-depth examination of both financial and non-financial aspects of an investment opportunity—in selecting PE-backed firms. Compared to traditional due diligence, ESG due diligence analyses risks and opportunities that are usually ignored or overlooked and, therefore, enables a more rigorous assessment of a company and its potential for improvement. In the fifth section, we focus on ESG factors in hedge fund strategies. Although the speculative nature of hedge funds may seem at odds with a socially responsible and environmentally oriented policy, ESG factors are often linked directly with notions of investor activism and good governance. In this respect, the monitoring and development of governance assessment programmes by hedge funds from an ESG perspective can add substantial value to investors' overall governance monitoring efforts. In the last section, we examine the ESG theme in real estate investments. Real estate management needs to take full advantage of recently acquired experience, knowledge and potential opportunities, and capitalise on these in its policies; this includes an analysis of ESG issues as part of the investment evaluation and decision-making process, identified as an outperformance opportunity.

2.2 Alternative Investments: An Overview

The term "alternative investment" has come about as a reference to the way in which capital is used: it implies an approach that is not the same as traditional investment methods. Broadly speaking, alternative investments refer to a large number of asset classes that are not part of traditional investments, such as equities, bonds and cash (Baker and Filbeck 2013). The emergence of this type of investment is rooted in sophisticated investors seeing the traditional investment environment as too low-return and looking for a way to achieve their return targets and, perhaps to a lesser extent, as a means of controlling risk (Chen et al. 2002; Anson et al. 2010). That is, they have seen alternative investments as an opportunity to achieve a reasonable return with manageable risk; some of these investments offer good opportunities to participate in different markets and to apply investment strategies that are not available to the general investing public (Agarwal and Narayan 2004; Jondeau and Rockinger 2006). As such, investors and portfolio managers who understand alternative investments wield a substantial advantage over those who do not.[1] Many previous studies have indeed found evidence of superior performance where alternative investments are concerned, whether on a stand-alone basis or as part of a portfolio of traditional assets (Chen et al. 2002; Amin and Kat 2003; Chen et al. 2005; Anson et al. 2010). For example, Anson et al. (2010) argue that alternative investments can obtain attractive rates of return, even in unfavourable market conditions. Ennis and Sebastian (2005), Lee and Stevenson (2005), and Bessler and Wolff (2015) all conclude that diversification into alternative assets is beneficial. Sa-Aadu et al. (2010) and Jackwerth and Slavutskaya (2016), too, find that diversification into alternative investments improves performance. This could also be attributed to the fact that a feature of alternative investment strategies is to use different means to profit from market volatility: for example, going short when the price is high and covering the position when the price is lower. By means of shorting, a company—and sometimes the entire economy—gains when the market falls. However, shorting is a high-risk activity because stock prices can recover unexpectedly and hedging short positions with leverage can become very expensive.

As alternative investments are a broad category, it is useful to limit the discussion to a few main types. According to Yau et al. (2007), we can divide these investments into two main categories—traditional and modern—which we summarise in Table 2.1.

Among the traditional alternative investments, we find commodities, private equity and real estate. Commodities are the name given to an agreement for the purchase and sale of a tangible asset or a real physical asset, generally of a relatively homogeneous nature. The three main commodities classes are energy (e.g. crude oil and coal), metals (e.g.

[1] Generally, a fund manager tries to achieve maximum performance by buying a security when it is cheap and selling it when the price has gone up. His/her goal in this case is to outperform the market index.

Table 2.1 Alternative investments categories

Traditional alternative investments	*Modern alternative investments*
Commodities: an alternative investment that consists of selling or buying a tangible asset or an actual physical good that is generally relatively homogeneous in nature	*Distressed securities*: financial instruments issued by a company that are either already in default, under bankruptcy protection, or in distress and heading towards such a condition
Private equity: an alternative investment class that consists in the provision of capital to non-listed companies on a public exchange	*Hedge funds*: an alternative investment that uses pooled funds that employ different strategies to earn an active return for their investors
Real estate: purchase of property as an investment to generate income rather than using it as a primary residence	*Managed futures*: an investment in which a portfolio of futures contracts is actively managed by funds and institutional investors to provide market diversification

Source Authors' own elaboration

gold, silver, platinum and aluminium) and agricultural products (e.g. coffee, beans, sugar and wheat). Private equity (PE) is an alternative form of private financing in which funds and investors invest directly in unlisted companies with potential growth. It involves a range of investment activities, but the most important areas of PE activity include venture capital (equity financing of new or growing private companies), closely controlled companies and buyout funds (the buyout of established companies through PE funds). Real estate is the ownership of an interest in land or structures attached to land. Investors may participate directly or indirectly, with direct ownership involving investment in residences, commercial real estate and agricultural land, and indirect investment meaning, for example, investing in companies engaged in: real estate ownership, development or management; real estate investment trusts; commingled real estate trusts; and infrastructure funds.

Among the modern alternative investments, we can distinguish between distressed securities, hedge funds and managed futures. Distressed securities refer to securities of companies or government agencies that are currently, or close to being, bankrupt. The most common distressed securities are bonds and bank debt. They are a high-risk option because, of course, the company may not recover and go bankrupt. Hedge funds are freely regulated and actively managed collective investment vehicles that employ a wide variety of strategies, the most common being aggressive long and short positions and the use of arbitrage and leverage. Since hedge funds can take many forms, there is no precise or universally accepted legal definition. However, their main purpose in most cases is to reduce volatility and thus risk by seeking to preserve capital and provide positive returns under all market conditions. Finally, managed futures refer to private pool investment vehicles that can invest in cash, spot and derivative markets for the benefit of their investors with the ability to use leverage in a wide variety of trading strategies. They also have the potential for reducing portfolio volatility and can earn a profit in any economic environment. Managed futures accounts can take long and short positions in forward contracts and options on forward contracts in global stock, commodity, currency and interest rate markets.

Although each of these alternative investments has unique features requiring a different approach from investors, they also have some features in common. For example, they can all be relatively illiquid and may incur relatively high purchase and sale costs compared to equities and bonds. Assessing their performance is often difficult because of problems associated with determining the current market value of the asset in question and the complexity of establishing valid benchmarks. Also, the data may be too limited to allow a proper valuation of the investment. In addition, many alternative investments may not be available due to their complexity or structure. Complexity is a limiting factor for the average investor because such investments may require a due diligence process and a high degree of investment analysis prior to purchase. Structure refers to how the investment is offered, and many alternative investments are private offerings available only to sophisticated investors. In fact, alternative investments are mainly taken up by high-net-worth individuals (accredited investors) and institutional investors.

2.3 ESG Factors and Alternative Investments

The term "environmental, social and governance" (ESG) is often used to describe any style of investment that has a social purpose, although ESG aspects vary by country, industry and company. Under this umbrella, we can find concepts such as capital management, accountability, transparency, quality of leadership, strategy and execution, competitive dynamics in the industry, human capital management and efficient use of resources. Another frequent term used for an investment with a social purpose is "socially responsible investment" (SRI), which focuses on the impact of companies operating in specific areas of interest and will, for example, exclude from its investment strategies those in sectors deemed unacceptable, such as tobacco, alcohol, firearms or gambling. Countries as a whole might be excluded if they have records of human rights violations, which has affected companies active in apartheid-era South Africa and, more recently, those engaged in trading in minerals from conflict zones. As well as an exclusion approach, SRI can positively favour investments in companies or organisations whose activities are oriented towards social justice or environmental solutions, or provide services to a local community. The key distinction between the two terms "SRI" and "ESG" is that the former concerns the screening of companies while the latter provides guidance on company conduct with regard to its governance structures. Responsible investment has captured the interest of many institutional investors, based on a number of factors including a recognition by the financial community that:

- ESG factors play an important role in determining risk and return;
- An integration of ESG factors is part of investors' fiduciary duty to their clients and beneficiaries;
- Differentiation by offering responsible investment services is a potential competitive advantage;
- In a globalised, social-media world, issues such as climate change, pollution, working conditions, employee diversity, corruption and aggressive tax strategies pose genuine value-destroying reputational risk.

A responsible investment therefore differs from other investment strategies because it emphasises a moral and ethical return. As such, ESG factors can have a positive effect on corporate clients and it is therefore right and proper that they should not be ignored when making investment decisions.

Companies are increasingly including ESG responsibilities in their strategic planning, reporting and day-to-day operations, which contributes to the growing support they are receiving from institutional investors involved in alternative investments, who show a growing preference for ESG factors. Relevant to this is identification by the Bank of England's Prudential Regulation Authority (PRA) of climate change as a financial risk pertaining to the objectives of the bank and other institutional and corporate investors (PRA 2018). Key asset managers are vigorously committed to offering their clients a range of ESG investment choices. BlackRock, for example, suggests identifying and investing in companies that have an impact on social and environmental issues and therefore score high on ESG factors.[2]

The main reason why many investors are focusing on social value is the growing appreciation that investments with negative ESG factors have an increased risk of negative financial performance, especially over the long term (e.g. Sassen et al. 2016). In fact, despite a number of conflicting studies, there is general support for the idea that companies applying ESG criteria achieve very high financial performance. For example, Clark et al. (2014) found that 90% of the over 200 studies they examined show that robust sustainability standards reduce the cost of capital; 88% show that companies with strong ESG performance enjoy better operating performance with high and stable cash flows; and 80% show that companies with strong sustainability and ESG practices score highly even against traditional financial metrics.

A paper by Fulton et al. (2012) argues that sustainable investment can be of great benefit to investors and companies and that ESG analysis, because it increases shareholder value, should be integrated into the investment processes of every serious investor and into the corporate strategy of every company. In an examination of over 100 academic studies, 56 research papers, two literature reviews and four meta-studies, the authors find all studies to be in agreement that companies with high ESG ratings enjoy a lower cost of capital in terms of debt and equity. In addition, 89% of the studies analysed link companies with high ESG ratings to positive performance based on market variables, while 85% show such companies enjoying positive performance based on accounting variables.

Friede et al. (2015) examined around 2,200 academic studies that analysed the relationship between ESG and financial performance. Their results support the ESG investment business case. Of those studies, 90% found a non-negative relationship between ESG and financial performance with the vast majority of studies finding it positive. They also find

[2] Larry Fink, President and CEO of BlackRock—the world's largest asset manager with over $6 trillion of assets under management (AUM)—has publicly challenged corporate CEOs to manage their companies with greater emphasis on ESG issues. https://www.blackrock.com/corporate/investor-relations/larry-fink-ceo-letter.

that governance shows the strongest correlation with positive financial performance. The authors' conclusion is that ESG factors may uncover non-financial risks that can become costly for companies in the future.

A common goal of ESG investments is to influence companies to behave consistently with investors' values around environment, society and governance. One way to do this is to interact with boards of directors and management. Simply divesting one's shares, however, does not always have the desired effect, for two main reasons. First, if all the concerned shareholders sold their holdings, the remaining (presumably less concerned) shareholders would be less likely to push an ESG agenda. Second, institutional investors may not have an incentive to do this because of the costs associated with both divestment and the diversion of personal resources.

Gond et al. (2018) examine the gains made from ESG commitments for both investors and their investee companies. Their analysis is based on 36 interviews with large listed companies, combined with previous research into the engagement practices of 66 institutional investors. While most previous studies show ESG engagement contributing to improved returns, this study focuses on how ESG engagement creates value for both investors and companies. To do this, it adopted multiple definitions of "value": "communication value", which means greater exchange of information; "learning value", which means the production and dissemination of new knowledge about ESG factors; and "policy value", which means the policy benefits that can be derived from commitment—for example, through increased executive support for ESG issues. These concepts are summarised in Table 2.2.

Gond et al.'s study provides key takeaways for both companies and investors. In particular, companies can improve communication with investors by more closely coordinating internal ESG information systems, ESG engagement information and ESG reporting practices. In addition, companies can benefit by acting more proactively, rather than responding only to investors, with regard to ESG policies and management systems. Finally, companies can also derive benefit from better internal coordination between investor relations, sustainability departments and board-level executives before meeting with external investors.

Takeaways for investors are the benefits of improving transparency, the communication of their engagement goals, and how their engagement processes are initiated, executed and evaluated. In addition, learning value can be augmented by paying more attention to ESG information

Table 2.2 Overview of each of the three value creation streams

Value creation dynamics	*Corporations*	*Investors*
Communicative *Exchanging information*	Clarifying expectations and enhancing accountability	Signalling and defining ESG expectations
	Managing impressions and rebalancing misrepresentations	Seeking detailed and accurate corporate information
	Specifying the business context	Enhancing investor ESG communication and accountability
Learning *Producing and diffusing knowledge*	Anticipating and detecting new trends related to ESG	Building new ESG knowledge
	Gathering feedback, benchmarking and gap spotting	Contextualising investment decisions
	Developing knowledge of ESG issues	Identifying and diffusing industry best practice
Political *Deriving political benefits*	Enrolling internal experts	Advancing internal collaboration and ESG integration
	Elevating sustainability and securing resources	Meeting client expectations
	Enhancing the loyalty of long-term investors	Building long-term relationships

Source Authors' own elaboration

and insights gained from engagement and active participation rather than relying on outsourcing for engagement activities.

Finally, takeaways for other business stakeholders are the benefits of working closely with ESG staff, and involving clients and beneficiaries in developing and refining engagement policies and objectives.

Some of the largest public and private investment managers dealing with alternative investments have become more proactive in propelling investee companies towards ESG practices by means of discussions with managers and boards of directors. BlackRock, for example, engages with investee companies in a constructive manner by asking questions rather than being prescriptive. If shortcomings are discovered in governance practices, dialogue will be entered into and BlackRock will expect a response consistent with its investors' long-term perspectives. BlackRock established several key commitment priorities for the two-year period

2017–2018, which include ESG policies, underlining its commitment to making a positive contribution to society in the areas of climate change,[3] human capital management, diversity and executive remuneration (Fink 2018).

CalPERS, the California civil servant pension scheme, also sees itself as an investor with a long-term investment horizon and, as such, considers ESG factors in all its day-to-day investment activities.[4] Indeed, it believes that long-term value derives from the effective management of financial, physical and human capital, a belief that underpins its sustainable investment agenda. There are several specific facets to CalPERS's corporate engagement programme. The first is based on environmental practices, including but not limited to climate change and the availability of natural resources. The second is based on governance practices, including but not limited to the alignment of interests. The third refers to human capital practices, including but not limited to fair labour practices, health and safety, responsible contracting and diversity. In addition, CalPERS has formed an alliance with other like-minded major institutional investors to make a commitment to their investee companies to adhere to responsible investment practices. These other investors include California State Teachers' Retirement System (CalSTRS), Connecticut Retirement Plans and Trust Funds, Florida State Board of Administration, Maine Public Employees Retirement System, Maryland State Retirement and Pension System, Nuveen (the asset manager of TIAA), OIP Investment Trust, Oregon Public Employees' Retirement Fund, Rockefeller Asset Management, San Francisco Employees' Retirement System, State Street Global Advisors and Wespath Investment Management. The alliance undertakes to monitor target companies' performance over time and to engage with them regularly. It is important that the key investor principles are reasonable and non-prescriptive in nature, and that the target companies are free to apply them as they see fit.

A specific purpose of this group of institutional investors is to create and promote principles for a responsible firearms industry, based on the belief that companies involved in the production, distribution and sale of

[3] Blackrock's specific goal in this regard is to improve the understanding of the processes that each company has in place to manage climate risks.

[4] CalPERS is the largest public pension fund in the US. It has over $300 billion invested in more than 10,000 public companies.

firearms are well placed to support the pragmatic transparency and security measures that can contribute to their responsible use. Although the group recognises that the civilian firearms industry is highly regulated in terms of production, sale and use, it sees the opportunity for improvement through voluntary commitment by the companies in this sector. It therefore calls on civil firearms companies to publicly demonstrate and publish their compliance with each of these five principles, ensuring that they adopt all necessary instruments to mitigate the risks.

- Principle 1: Manufacturers should support, advance and integrate the development of technology designed to make civilian firearms safer, more secure, and easier to trace.
- Principle 2: Manufacturers should adopt and follow responsible business practices that establish and enforce responsible dealer standards and promote training and education programmes for owners designed around firearms safety.
- Principle 3: Civilian firearms distributors, dealers and retailers should establish, promote and follow best practices to ensure that no firearm is sold without a completed background check in order to prevent sales to persons prohibited from buying firearms or those too dangerous to possess them.
- Principle 4: Civilian firearms distributors, dealers and retailers should educate and train their employees to better recognise and effectively monitor irregularities at the point of sale, to record all firearm sales, to audit firearms inventory on a regular basis, and to proactively assist law enforcement.
- Principle 5: Participants in the civilian firearms industry should work collaboratively, communicate and engage with the signatories of these Principles to design, adopt and disclose measures and metrics demonstrating both best practices and their commitment to promoting these Principles.

This framework affirms the group's role as proactive investors and identifies expectations for the firearms industry that will reduce risks and improve the security of civil society in general.

2.3.1 The Role of Impact Investment

One of the challenges of recent decades has been to integrate impact and impact measurement into investment decision-making using management systems, multi-criteria decision analysis and the application of systemic approaches.

It is mainly institutional investors who are currently driving the impact assessment process and its integration into investment and finance. There are several reasons for this, including effective risk management, reputation protection and the fulfilment of stakeholder requirements. The World Bank, international financial institutions, other multilateral bodies and the private banking sector have been collaborating in pursuit of relatively consistent global ESG standards, a process often referred to as "global administrative law" (McIntyre 2015)—a mix of voluntary and regulatory initiatives. In fact, most institutions have begun to adopt global ESG policies and standards as part of their alternative investment choices to circumscribe the projects and activities they finance. This is especially the case for the financing of large infrastructure projects in developing countries or economies in transition (McIntyre 2015).

There are several good reasons for institutional investors to integrate ESG factors into alternative investment decisions, which include reputation, fiduciary duty, risk management and, last but not least, the emergence of the aforementioned "global administrative law". In fact, several studies have shown ESG factors to have an inherent potential to create financial innovation based on environmental, social, technological and organisational policy analysis.

According to the Rockefeller Foundation (2007), impact investments are those made in companies, organisations and funds with the intention of generating a positive social and environmental impact along with a financial return. They can be made in both emerging and developed markets and across asset classes, including bonds, listed equities and private equity. Such investment will include ESG criteria and will have a long-term investment view (Harji and Hebb 2010; Harji and Jackson 2012; Harji et al. 2014; JP Morgan Social Finance 2014; Wilson 2014). Impact investments can provide a comprehensive assessment of a company's organisational performance based on three main pillars: first, the external legal and administrative, political, cultural and economic environment; second, organisational motivation, i.e. history, mission, awards and incentives; and third, organisational capacity, i.e. strategic leadership, human resources, programme management and financial management. The three pillars combine to provide an overall assessment of the organisation's effectiveness, efficiency and financial sustainability.

Some studies (e.g. Cambridge Associates 2015; New Philanthropy Capital 2015) have identified a journey that investors have undertaken, which begins by going from responsible investment to sustainable investment; the latter is understood, by the majority of industry stakeholders as sustainable implementation of management practices in relation to ESG criteria. The ESG factors are then transformed into a driver of innovation, becoming a catalyst, while the core ESG values are maintained. In other words, the value creation process leads to a thematic investment strategy and ultimately to an impact-based investment strategy (Cambridge Associates 2015; New Philanthropy Capital 2015). This journey is summarised in Fig. 2.1.

Impact investing should also be understood as a proactive process through which investors select, assess and monitor investments using ESG criteria. While a responsible investment—or socially responsible investment (SRI)—seeks to avoid the risks associated with socially and environmentally damaging investments, impact investing aims to actively

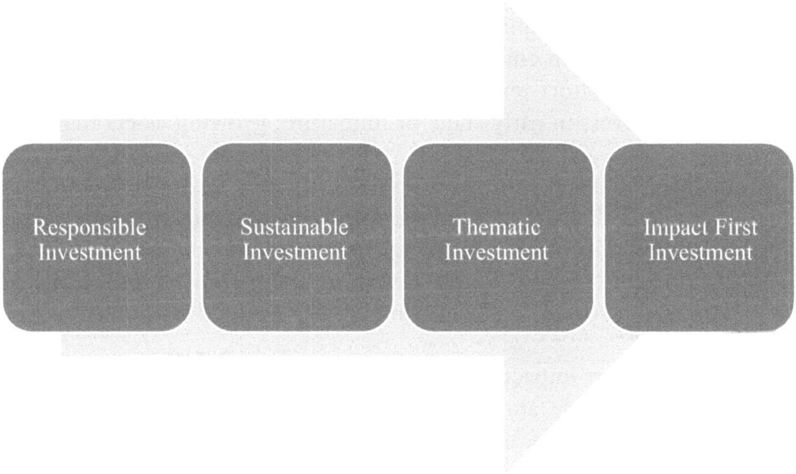

Fig. 2.1 The impact investment journey (*Source* Cambridge Associates [2015] and New Philanthropic Capital [2015])

and intentionally create a positive and measurable impact through profitable activities for the target company by systematically applying ESG criteria against risky assets. Thus, investors will adopt leadership positions and seek out ESG gaps by analysing the corporate culture using due diligence and gap analysis. This begins with a comprehensive gap report which includes ESG, leadership and cultural factors, after which the gaps are actively addressed. Thus, influence is first brought to bear on the leadership of a target company before the investment is made, and then exercised by the investors, as active owners, throughout the life cycle of their investment.

2.3.2 Vehicles for Impact Investments

Investors use a variety of vehicles to activate impact investments, and this includes the creation of private equity or venture capital funds, using direct investment strategies. JP Morgan Social Finance and the Global Impact Investing Network (GIIN) have shown that private equity is in fact by far the most commonly used tool for impact investment (O'Donohoe et al. 2010; Saltuk et al. 2014). These two organisations have further explored the dynamics of impact investment in various publications, such as *Perspectives on Progress: the Impact Investor Survey*, and they show that private equity and venture capital can look at longer time horizons and therefore embrace longer-term patient money while being less dependent on short-term outcomes. In addition, both these types of funding often invest in early-stage or immature, growing assets that have not managed to reach critical scale.

Companies involved in impact investments may be startups and may occasionally find themselves funding entrepreneurs also in business startups. Both discipline and risk tolerance is therefore required in order to prepare such companies and invest in their restructured activities while imposing a socially and environmentally oriented corporate culture (Wendt 2018).

The research on impact investing is still sparse, especially compared to that on responsible and sustainable investment (Clark et al. 2014). Furthermore, although academia defines impact investment as comprising responsible and sustainable investment, impact investors themselves clearly distinguish between the two concepts (Brandstetter and Lehner 2015). In fact, impact investments do not follow the logic of traditional financial instruments (Brandstetter and Lehner 2015). Because impact investing explicitly seeks to generate quantifiable social and financial returns, finding a generally accepted way of measuring such returns will be a key component of necessary future research.

According to the World Economic Forum, the major asset owners currently allocating capital for impact investing include financial institutions for development, private equity and venture capital funds, family offices and high-net-worth individuals. However, these types of investors hold only a small share of the global capital pool (World Economic Forum 2013). Therefore, an enquiry into what factors prevent other types of asset owners from allocating capital to impact investment would be an important exercise.

It is also important to understand that impact investment is an investment approach and not an asset class: it is a criterion according to which investments are made through asset classes. They belong in a category of investments motivated by the intention to create a social or environmental asset. It is an approach that measures financial returns in relation to social and environmental impact (World Economic Forum 2013).

In order to understand the structure of impact investment, Bridges Ventures, Lehner and Brandstetter have developed a method of classifying different forms of impact investments. Figure 2.2 shows the structure of the impact, from responsible to philanthropic, and presents a classification

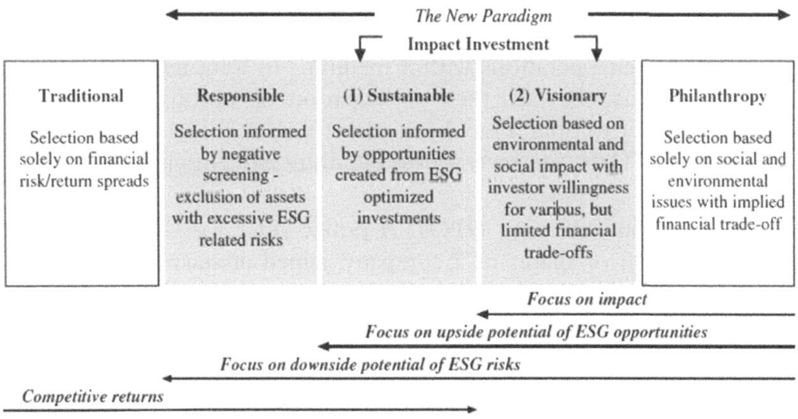

Fig. 2.2 The spectrum of capital (*Source* Nicklin [2012], Clara Barby, Bridges Ventures)

scheme for the various forms of impact investing. Practitioners describe a journey from responsible investment to impact investing, focusing primarily on thematic issues that create added value for the client, whereas the academic world describes a path taken through the pillars of the responsible and sustainable investment process to "visionary" and "philanthropic", which differs from the practitioners' definition in that it focuses more on social issues.

2.4 Private Equity

Private equity (PE) is an investment approach that aims to generate a substantial financial return for fund managers and their investors. PE funds are structured such that the limited partners contribute the capital and the general partner allocates the investments. In this regard, venture capital can be considered a subset of PE investments, the distinction being that venture capital is usually invested in startup or early-stage companies, while PE typically invests in mature companies. PE companies have also been associated with strategies for purchasing companies using leverage or high debt levels. This is generally followed by the restructuring of the company and the sale of the resulting entity to partners at a large profit. The process of a PE transaction looks like this: investors, through an investment vehicle—i.e. a PE fund—raise capital from major institutional and individual investors (such as banks, pension funds and high-net-worth individuals), committing mixed capital in private companies to expand and improve their operations with a medium- to long-term perspective; after several years, they sell their participation via various disinvestment channels in order to offer a rapid exit path for shareholders and prepare a liquidity event. There are several possible disinvestment channels:

- Initial Public Offering (IPO). A public offer for the purchase or subscription of shares of a company, aimed at starting the process of listing on the stock exchange.
- Trade sale. A disinvestment method through which the PE investor's stake is sold to an industrial purchaser, determined on the basis of a strategic motivation.
- Buy-back. Selling the PE investor's stake back to the entrepreneur who originally sold the shares.
- Secondary sale. Selling the PE investor's stake to another PE fund.

Generally, the life cycle of a PE transaction can be divided into four main phases. The first consists of raising funds, then seeking capital from investors. In the second phase, a PE company will look for investment opportunities, i.e. procurement. Usually, PE firms find all the useful information about target companies during the third phase: due diligence. As soon as a PE firm has identified a target company and concluded a transaction, the third phase begins, i.e. managing the companies in the portfolio. PE firms can increase the value of the companies in their portfolio through a variety of different transactions, such as organically expanding a business by making acquisitions in the same sector and improving the market share of the target company. The last phase, "exit", consists in selling the shareholding of the target company in order to earn a return on the investment.

The relationship between PE and ESG, as explained in the literature, dates back to 2009, when the US Private Equity Council established a responsible investment policy by adopting guidelines on environment, health, labour, governance and social issues (Schell 2020). Recent studies have revealed that, initially, ethical investment criteria included negative screens to exclude companies in certain sectors or industries for moral or ethical reasons (Crifo and Forget 2012). Over time, as concerns over negative screening emerged, ethical and socially responsible investments evolved which integrated ESG factors. According to Cappucci (2018), the integration of ESG factors into investment strategies addresses the problem of those managers who adopt ESG being disadvantaged when competing with managers who have no ESG costs to bear. Although in recent years PE and ESG have quickly become strongly associated with each other, the literature is yet to offer any clear analysis of the integration of ESG aspects by PE firms: for example, which valuation tools they prefer, which criteria they use to perform ESG due diligence, and what main barriers they face in integrating ESG.

Finally, when we talk about PE in relation to ESG issues, we need to distinguish between funds that use these factors as selection criteria for their target companies and funds that use the impact investment approach.

2.4.1 ESG Selection Criteria in Private Equity

PE, as a subset of social investment strategies between companies and investors, has various responsible investment strategies for selecting target companies. The three main ones are (Blanc et al. 2009; Wood and Hoff 2007):

- A product or a thematic approach;
- Economically targeted investment or community enterprise;
- Process-focused investment or ESG screening.

PE's relationship with ESG is that it incorporates these criteria into the investment decision-making process and also helps in assessing the ESG information required by investors when selecting initiatives to invest in. Similarly, PE firms themselves can be assessed using these approaches to determine whether ESG aspects are taken into account during the investment decision-making process. PE firms may also select companies whose products and services offer sustainable solutions appropriate to the needs of the target company; this is the thematic approach.

By the nature of these activities, PE investors can therefore provide their target companies with support in developing environmentally and socially beneficial products and services (Blanc et al. 2009, p. 9; Wood and Hoff 2007). In fact, common criteria on PE investors' radars are environment, demographics and welfare services, sustainable agriculture and fair trade. Here, the "environment" component might consist of climate change, energy efficiency, renewable energy and waste and water management, among others. An example of a category of investments focused on green products, renewable energy and health innovations goes under the umbrella of "Cleantech". The demographic and welfare services offered by target companies can be supported by investments affecting a particular geographical area of activity as well as the quality of the service or product offered (Blanc et al. 2009, p. 9; Wood and Hoff 2007). An example is the US company Braemar Energy Ventures, which invests in energy technology companies that provide environmental solutions such as clean fuel processes, pollution control and advanced energy generation.

According to Wood and Hoff (2007), another characteristic of PE companies is that they make new technologies and techniques available to target companies; it is therefore more likely that investors through their investments and through the use of these technologies can lead to a positive impact on sustainability. In addition, PE firms use, *ex ante*, a responsible investment strategy for selecting target companies which is similar to the responsible investment screening strategy between large capitalisation companies. This is achieved by integrating companies' analysis of ESG asset management into the investment decision-making process. At this stage, investors may decide to focus only on companies with good governance or good environmental management. In addition, the screening strategy may also include companies that promote sustainability or culturally based initiatives, or exclude companies that engage in activities that are considered unethical, e.g. arms (Blanc et al. 2009; Wood and Hoff 2007). Through these investment strategies, PE firms can mitigate the risks associated with the funds invested by considering non-financial issues.

By its nature, the PE industry can influence the management decisions of the target companies it invests in. The opportunity to sit on the board of directors allows them to incorporate ESG aspects into their portfolio activities, making it easier to influence those companies' corporate cultures in the direction of social responsibility and sustainability (Wood and Hoff 2007). Wood and Hoff (2007) provide an overview of recommended considerations for investors in selecting fund managers to manage their process-focused investment portfolio:

- Establish the key ESG performance indicators against which companies and fund managers should be measured. At this stage, it is critical to ensure that the information on which these indicators are based is actually available. For example, evidence of environmental performance might rely on regulatory compliance issues; workplace performance might use indicators such as the basic salary in other companies of similar size and in similar industries.
- Check that the fund manager implements company-specific strategies to improve the management of ESG aspects prior to investment, and then requires the target companies to measure these factors against business objectives.
- Check that the fund manager provides technical assistance to the target companies to help them achieve greater integration of ESG risks and key opportunities related to their business.

In addition to this, Steffensen (2006) identifies PE investors' need to implement ESG criteria as being driven by high risks and investment opportunities. Some market participants, such as financial analysts, regulators and investors, share this concern and believe that investment decisions and corporate assessments could be improved were they to adequately reflect their related ESG risks (Amaeshi and Grayson 2008). ESG aspects are already influencing the business environment in areas such as investments, green and sustainable funds, and the more commonly identified stakeholders, such as consumers (Miles 2010).

The ESG criteria need to be effective and efficient, and this requires a well-defined strategy that includes ESG issues from the beginning to the end of a PE investment decision-making process. In this regard, ESG research can provide a long-term view of target companies' prospects, which can help to identify and exploit market opportunities to maximise returns for investors (Dixon 2009).

In integrating ESG aspects, the pre-investment phase is important for PE companies: ESG decisions in this phase can be crucial in ensuring effective monitoring in the post-investment period. In fact, ESG integration can begin as early as the creation and selection phase of the transaction. At this stage, a PE investor with an ESG focus will evaluate potential initiatives to determine whether the proposed agreement will lead to synergies, opportunities or conflicts in relation to their investment policies. Responsible PE investors may therefore ultimately decide to invest in clean technology initiatives (or "Cleantech"), renewable energy or other innovations based on ESG issues (PwC and Waterman 2010).

To summarise, at this stage, a PE investor should:

- Assess whether there are significant ESG risks or opportunities related to the industry sector or geographical location of the company;
- Determine whether the potential ESG risk capital policy adequately addresses any significant concerns that may impact the industry;
- Assess whether there are any potential short-term impacts on the target company's business model in the light of ESG factors.

During these phases, careful and effective ESG due diligence is crucial in verifying the feasibility of the operation. PE firms undertake due diligence in order to understand the target company's performance and assess any risks or opportunities that could impact the business model. In addition, ESG aspects should be considered with regard to both legal compliance and non-regulatory issues (PwC and Waterman 2010).

In the post-investment phase, the PE investor becomes an integral part of the new management, where he or she can monitor and support the implementation of ESG factors. This phase therefore facilitates the encouragement and motivation of responsible business practices in the companies in which they invest. Several benefits derive from this, such as the protection of financial interests in the company, effective monitoring and support with the objective of protecting or increasing the company's profitability, and increasing company value during the exit phase (PwC and Waterman 2010).

2.4.2 *Impact Investment Criteria in Private Equity*

Impact investors use a number of criteria in their investment strategies that distinguish them from conventional investors. First and foremost, they must consistently define and measure impact. While many PE investors in emerging markets generate a positive economic impact through their investments, this alone does not qualify them as impact investors. To qualify as such, these funds must define, analyse, integrate and manage the impact throughout the life cycle of an investment. They must also demonstrate that impact considerations are integrated throughout the investment process, from initial screening, due diligence (including ESG aspects), to closing and post-investment monitoring with measurable results. Adherence to all these criteria will allow them to be differentiated from purely financial PE funds by intention, measurement and accountability.

In summary, therefore, once a PE impact fund has made an investment, it continues to closely monitor its impact. Funds generally interact with target companies on a quarterly basis, monitoring performance using metrics that will vary according to sector. There are many organisations trying to introduce standardised metrics, and these include Impact Reporting and Investment Standards (IRIS) and Global Impact Investing Ratings System (GIIRS), but as yet there is no universally accepted approach. So it is important that the fund specifies, *ex ante*, the relevant metrics and is held accountable for them.

In pursuit of increased effectiveness, many PE impact funds create a value statement and change theory against which they measure their investment.[5] According to TriLinc Global, by having such a social mission as part of the investment strategies, the integration of intended impact with financial objectives confers on funds the following benefits:

[5] The theory of change highlighted by Jackson (2013) is used to measure impact analysis, as it aims to build and address the measurement of the impact on the social and environmental aspects of investments.

- Integration of monitoring and analysis of data on both finance and impact performance;
- Formalisation of investors' responsibility for the impact;
- Mitigation of the potential trade-off between performance and impact.

These are of benefit to the target company's business because they improve its management, corporate governance and legal compliance, thereby creating competitive advantage. One of the hallmarks of PE funds is the imposition of discipline and efficient execution in fast-growing companies. As mentioned above, PE operators' role in educating target companies can be a remarkably effective one in those companies that are looking to implement new business models, which imply volatile cash flows (e.g. Metrick and Yasuda 2010; Michelini and Fiorentino 2012).

PE impact investors can therefore make good use of a sustainability and social policy, changing companies' business models and improving the way they do business, by bringing innovation and offering capital (Loew et al. 2009; Porter and Kramer 2006; Moore et al. 2012). In fact, any growing PE equity fund that invests in companies that meet ESG criteria, and not only in emerging markets, is looking to transfer management and operational expertise to portfolio companies.

Companies' difficulties in attracting institutional capital remain a significant constraint on the development of impact investing. Although it has grown in size and importance in recent years, PE impact investing remains a rare strategy, according to Bridges Ventures, and not one that traditional institutional investors generally include in their investment strategies. To attract more institutional investors, there is a need to demonstrate that it is possible to achieve both impact and financial returns and to educate investors about investment opportunities. FIR Capital has sought to address this gap in Brazil by convening private asset managers, the Brazilian private equity association, pension funds, universities and journalists, with the support of the Brazilian private equity association ABVCAP (Wendt 2015).

Relevant and robust metrics are also of key importance during the investment period in order to demonstrate success in achieving social and environmental impact. However, the idiosyncratic nature of impact investing presents at least two specific challenges with respect to the development of such metrics. The first is the time scale: while financial returns for PE investors end once the fund has exited the investment, the social impact must continue after the investment period. Some projects create an impact throughout the life of the investment, while others have a longer-term impact—but in many cases only from the final stage of the investment. According to Vital Capital, projects with an immediate impact should be differentiated from those with a long-term impact and measured differently. The second challenge concerns the value differentiation between results and outputs. Results, such as poverty reduction, reflect the final impact target of impact investments, while output uses metrics such as number of housing units built. Moreover, results are more difficult to measure, as identifying the causal link between a company's operations and the result is an expensive process. Furthermore, it can be a challenge to attribute a given result to a particular investment.

Performance monitoring of target companies adopting ESG criteria is done using a series of standards, which cover, among other things, organisation, product and financial performance. This is carried out by Impact Reporting and Investment Standards (IRIS), managed by the Global Impact Investing Network (GIIN), a network founded in 2008 by impact investors, namely the Acumen Fund, B Lab and the Rockefeller Foundation. According to Wendt (2015), PE impact investors need to simplify this monitoring process and make it more practical by focusing on the most important metrics. Marcus Regueira of FIR Capital recommends just four to five indicators per industry in order to provide a balance between comparability and indicator overload.

2.5 PE Investments and ESG Due Diligence

As described above, the PE business model can be divided into four phases: screening; investing in the target company; adding value to the portfolio of target companies; and, finally, selling those companies. Due diligence, which involves the integration of ESG criteria, adds value to each of these steps and aims to increase target companies' performance. An example of a PE firm that adopts ESG criteria in its due diligence is Aloe Private Equity, which manages a number of environmental activities and socially sustainable funds and is dedicated to investing in companies that provide solutions to environmental and social problems while offering satisfactory financial returns to its investors.

Thus, from the moment a PE firm invests in a responsible company, it has begun to promote better corporate social governance, better environmental performance and safety reporting, as well as initiating the traditional corporate restructuring generally associated with PE. This full-spectrum approach, supported by legal rights, is critical in improving high-impact ESG aspects within the target companies.

In the due diligence phase, PE investors will assess whether ESG factors are a cost or a profit driver. If they are seen as a business cost, this may imply damage to the PE firm's reputation. The traditional view is that ESG assessment and management is seen as a cost, but that it serves to reduce business risk. But there is another view that sees the integration of ESG aspects as a profit driver leading to an improvement in the corporate culture with ESG-related benefits that lead to greater profitability. For example, environmental monitoring of air, solid waste and waste water informs business decisions regarding the loss of raw materials, products or production yields. Also, attention to social factors, such as health and safety, can reduce sick days and improve employee loyalty. And improvements in governance can reassure future buyers of a company and can lead to better business judgement when selling the company.

The due diligence of an investment by a PE firm consists of a series of decisions that start from the initial phase to the phase in which the final investment decision is made. For each phase, the required documentation and rationale becomes more thorough—and any ESG, financial, legal or recruitment issues raised in the previous phase must be resolved before the investment opportunity can move on to the next.

In a first step, the PE firm may exclude certain sectors from its investment choice. An assessment (high, medium or low) of the potential environmental and social impacts of the investment opportunity is made to ascertain the intensity of future ESG due diligence. During this phase, investment opportunities will more often than not reveal negative environmental and social impacts that are irredeemable, such as the loss of an important natural habitat, or a negative impact on ethnic minorities or cultural heritage. This is why a comprehensive, high-intensity assessment of the environmental and social impacts is crucial. High-risk ESG investment opportunities will include mining, hazardous waste disposal, oil and gas, and forestry.

Investment opportunities that reveal an average ESG risk might include general manufacturing facilities, food processing, paper mills and textile industries. But investment opportunities can also reveal negative or zero environmental and social impacts. In such cases, a social and environmental checklist, and an assessment of workers' rights and working conditions may be sufficient, providing no problems are identified. Investment opportunities with a low ESG risk might include office-based advisory services.

In the due diligence phase, PE firms also analyse potential development effects to assess the investment opportunity's benefits and determine whether it fits the sustainable investment objectives. Here, the required ESG standards for the target company are introduced. At a later stage, a social and environmental checklist, and one for labour rights and working conditions, are used to gather comprehensive information on the environmental and social aspects of the potential investment opportunity. This phase takes about a month, during which financial, human resources, sales and marketing operations are documented and analysed.

The results of previous analyses of social and environmental rights, labour rights and working conditions will dictate whether an environmental impact assessment is required. Given the complexity of assessing, for example, environmental discharge requirements or the potential severity of an incident, an independent expert may be asked to examine the ESG impacts of the potential investment opportunity. Usually, an independent firm of auditors will also be involved to assess the financial data, or an independent law firm to review the legal status. All findings and areas for improvement within all ESG aspects are integrated into the action plan, which will document any social and environmental shortcomings. Actions and time frames for remedying these deficiencies will be agreed in collaboration with the target company. Pre-closure and completion conditions will also be included in the action plan and in the legal documents. PE firms may reject an investment opportunity on the basis of irregularities in a company's ESG procedures as identified in the checklists, or else they will have to find ways of effectively managing problems that may affect investments. Often, a potential target company is required to sign a social and environmental code of conduct, approved by that company's board of directors.

Having completed this step as part of the due diligence process, and with the action plan agreed, the information will be incorporated into the legal documents upon which the PE firm may proceed with the final investment decision. The legal documents include ESG aspects and a corrective action plan in order to provide legal remedies in the event of non-compliance with ESG criteria.

The action plan and the agreed post-closure conditions form the basis of ESG management within the first year of investment. The PE investors will subsequently follow target companies' performance throughout the year by means of regular meetings and additionally require all portfolio companies to complete an annual monitoring report to document their company's ESG performance and their progress in implementing the action plan.

In this way, due diligence data allow PE investors to create an economic rationale based on net present value to demonstrate an improvement in the ESG capital spending line or ESG operational change. In addition, these data will describe the company's performance taking into account environmental performance with regard to air emissions, liquid and solid waste, and hazardous materials. This information is recorded with reference to national legal requirements. If emissions are in excess of national laws, immediate improvement will be required to justify capital investment and to dispel the risk of closure. Emissions measurements are also useful in providing data to understand the amount of raw material lost and the amount by which the yield of a process can be improved. Due diligence analysis also covers social performance, captured by number of employees, external collaborators, number of trade unions, male/female ratio, accident statistics, number of strikes and production losses. In this regard, investment in employees such as training and health and safety can be economically justified on the basis of reduced staff turnover and better relations with trade unions. This essentially means that an improvement in a company's social policies can increase profitability. Community relations are also assessed, i.e. activities in which the company has engaged with the local community, and complaints made by local communities.

PE firms use ESG data to provide a base from which continuous improvement over subsequent years can be measured. The process of annual monitoring, review and updating of the plan should be incorporated into the legal documents at the time of investment.

During exit negotiations, counterparties will assess the risks within a company to influence the assessment. Full compliance with ESG criteria—with supporting evidence—will enhance a company's financial value. As such, as PE investors reach the final stage—selling their stake in the company—a good ESG rating for the target company relative to the investment date is important in achieving a higher valuation and maximising capital gains.

Further, ESG reporting follows a model similar to the cash flow statement; in other words, results at the portfolio company level are consolidated at fund level, and the results of different funds are gathered in a set called the "fund of funds". The consolidated results are disclosed to the owners of the assets.

2.6 ESG Aspects in Hedge Fund Investments

A hedge fund is an aggressively managed portfolio of investments invested in a wide range of assets using leveraged, long, short and derivative positions. The structure of a hedge fund is similar to that of a PE fund, i.e. the capital is raised by investors, called "limited partners", and is invested by professional managers employed by the fund company, called "general partners". A hedge fund's investment strategy is to go long and short on the equity markets at the same time. The theory is: if the market rises, the long gains outweigh the short losses; but, if the market falls, the short gains outweigh the long losses. Therefore, the investor is at least partially covered by a fall in the stock market.

Hedge fund strategies have evolved over the past decade and many funds now invest in a variety of all the assets available, such as equities, debt, real estate, foreign currency and commodities. More recently, as with other investment managers, some hedge funds are beginning to offer ESG-sensitive investment funds. In fact, a recent survey of 80 hedge funds found that 40% of respondents said they make at least some investments using "responsible" investment principles. Around 10% of assets under management (AUM) were invested responsibly (AIMA 2018).

Given the multitude of investment strategies and options, it can be quite difficult for this category of investors to select a fund manager capable of generating a sustainable profit year on year. However, some investors may define governance to include social guidelines about the nature and type of investments that a hedge fund makes. An example of this is socially responsible investment (SRI). As described above (Sect. 2.2), in the SRI approach investment objectives are combined with certain ethical or environmental objectives (Sparkes 2002).

In a hedge fund context, this means that a fund manager should adhere to an investment policy with socially responsible objectives such as environmental protection (or should exclude others contrary to those objectives). The idea of a hedge fund having responsible objectives with an SRI orientation, considering their notoriously speculative nature, has led to scepticism. Nevertheless, supporters of SRI hedge funds argue that SRI funds can be profitable and effective and hedge funds seeking to join major SRIs simply need to comply with the SRI requirements (Hedge Funds Review Editorial 2013). Other than SRI policies, hedge fund investments might now also include ESG aspects related to corporate governance.

ESG factors are examples of the ways in which many investors seek to guide their hedge funds' investment strategies through permissive investment strategies. ESG criteria are often related to notions of investor activism and good governance—in part because, in addition to setting out these criteria, investors often seek to monitor how hedge funds implement them on an ongoing basis.

In this context, the shareholder activism commonly focuses on governance issues such as overcompensation of managers and misuse of corporate assets. Thus, although for large institutional investors ESG engagement strategies have evolved over time, where shareholder activism is concerned there has been little interest in social and environmental issues not directly related to increasing the value of equity in the short term. Also, the investor's ESG focus is typically geared towards encouraging investment in companies that are believed to be in line with their ESG values and avoiding those that are not. Investors will always ask the question, however: beyond the environmental benefits, is ESG investing actually profitable?

The literature reveals conflicting opinions on this. Some studies assert that ESG funds' track record is still too short to draw any meaningful conclusions about profitability (Staub-Bisang 2012). Rice et al. (2012) say that hedge funds cannot follow ESG policies if they want to be profitable. Röhrbein's (2012) finding is that hedge funds largely ignore ESG aspects in their investment strategies. Despite these conflicting voices in the literature, it remains the case that large institutional investors such as CalPERS adopt ESG criteria in their investment policies. And, according to Hua (2011), funds that have adopted ESG policies have also benefited by raising inflows from institutional investors. Notwithstanding these debates, it is clear that hedge funds seeking to adopt ESG and SRI criteria will agree to invest in companies only if their managers agree to follow certain guidelines. According to Lohr (2011), this could be a manifestation in hedge funds of wider public market efforts to promote social responsibility and good corporate citizenship. As discussed by Lindsay (2013), some groups of hedge funds have even decided to distribute premiums to funds that promote SRI investment approaches.

If an investor invests directly in a hedge fund with a sole objective of SRI, and which has its own predetermined SRI guidelines, he or she may choose, for example, to exclude oil companies for environmental reasons, and to invest instead in companies that promote cleaner energy, which is sometimes referred to as clean technology, such as solar energy or electric cars. Other funds have slightly nuanced differences, such as those oriented towards economic, social and environmental (ESE) and corporate social responsibility (CSR) principles. Then, there are investors who wish to take advantage of the principles of other parties such as the United Nations Principles for Responsible Investment (UNPRI). In the end, the critical point is the hedge fund's adherence to its set targets.

Ultimately, the objective of these initiatives is to promote what investors consider to be good corporate governance principles in the target companies and to generate substantial cash flows. Nonetheless, as noted above, these frameworks also have fund-wide implications. During the process of making an initial or ongoing governance assessment of a hedge fund, the ways in which funds are structured both to comply with these corporate governance guidelines and to monitor them in the future may be important indicators of the level of governance controls in place at fund level.

2.7 Environmental and Sustainable Real Estate Investments

Environmentally responsible and sustainable business activities have taken on an increasingly key role within companies' strategic and operational activities, as the focus on climate change and investment ethics increases. From an investor perspective, responsible investment covers not only environmental issues—for example, alleviating climate change—but also a wide range of social and ethical concerns. The real estate sector, as a case in point, has found ESG issues becoming consistently prominent: the unique contexts of each construction project, the large number of stakeholders involved in operations and the different financial pressures have all been identified as key challenges for the mitigation of unsustainable trends in that sector (Teo and Loosemore 2001). Ofori (1992), divided the impact of real estate activities on the environment into various categories: chemical pollution, deterioration of resources, visual impact and health impact. These categories subsequently became the established framework for sustainable real estate standards (Brandon and Lombardi 2010).

According to the World Economic Forum (2016), buildings consume over 40% of the global energy supply, and the real estate sector consumes about 30% of raw materials and 12% of freshwater globally, generating up to 40% of total landfill waste and 20% of water effluents. At individual country level, the situation is even more dramatic in the United States, where 40% of carbon dioxide emissions and 88% of freshwater consumption are related to real estate activities (Deloitte 2014). A study published by the Lawrence Livermore National Laboratory states that, in the US, 59% of total energy produced is wasted due to inefficient and ineffective use, with the construction sector and buildings identified as the main culprits (LLNL 2014). This situation is set to become even worse, given the increasingly urban nature of the world's population: the 750 largest cities in the world are expected to require 260 million new homes and 540 million square metres of new offices by 2030 (World Economic Forum 2016). Moreover, according to an estimate by the World Economic Forum (2016), CO_2 emissions from buildings and the proportional share of global greenhouse gas emissions will increase by 56 and 7% respectively by 2030. Given these concerns, scholars and policymakers insist that green buildings, sustainable development and environmentally friendly construction and management operations are of significant importance, and are seeking to push the real estate sector towards sustainable policies.

In this context, there are a number of analyses that substantiate the hypothesis that sustainable construction can improve energy and water efficiency and ensure the sustainable use of raw materials (UNEP 2011). In fact, according to The Impact (2016), a sustainable housing market could save 23.5 billion kilowatt-hours of energy between 2015 and 2018. In addition, the US Green Building Council and Booz Allen Hamilton (2009) state that these energy cost savings and climate benefits also bring financial benefits through labour cost savings, and productivity gains from increased indoor air quality and local thermal control. Other studies argue that increased daylighting and contact with nature also have positive effects in hospitals and schools by improving student performance and patient recovery (Ulrich 1984; UNEP 2011). It has been estimated that "green" buildings currently account for 38% of global construction project activities, highlighting the inroads that sustainable real estate activities are already making in the economic and social dimensions (Bernstein et al. 2013).

Bukart (2009) classifies sustainable economies into six macro-sectors: sustainable transport, water management, sustainable buildings, land management, waste management and renewable energy. As one of these six pillars, the construction sector can therefore play a key role in reducing negative effects on the environment—for example, through lower carbon emissions—and human health. Moreover, sustainable construction could also create job opportunities in the design and construction phases (UNEP 2011). Despite these benefits, the supply of sustainable buildings remains low, which some scholars blame on either market failures or structural problems within that specific market and industry that are related to asymmetric information and externalities. An example of construction processes leading to a negative externality is construction waste contaminating water and air (Matisoff et al. 2016). Real estate investments therefore need to look closely at available economic instruments that address these issues, such as energy efficiency contracts, cooperative procurement and credit schemes, and tax, i.e. subsidies or soft loans. One of the main opportunities for sustainable construction is the lower cost of modernisation or new construction when it comes to emissions reduction, compared to other economic sectors (UNEP 2011). In fact, it is estimated that sustainable construction practices could reduce emissions from new construction by almost a third at almost zero cost (IPCC 2007).

Sustainable urban development, which is fostering economic growth and innovation in many major cities around the world, has the potential to use resources more efficiently and effectively, protect the environment, create jobs, and provide a supportive environment for individuals and communities. This is a critical project, given that over five billion people—or 66% of the world's population—will live in cities by 2030, according to estimates by the World Bank (2016) and the United Nations Population Fund (2017).

Eizenberg and Jabareen (2017) argue that physical infrastructure has a major influence on social sustainability: done right, it can reduce environmental risks and improve community well-being. UNEP (2011) has highlighted the various benefits of sustainable real estate, which include an increase in workers' productivity and also their welfare, thanks to reductions in noise and air pollution. Several studies focus on the various types of sustainable urban strategies to explain how sustainable real estate can impact climate change risk management (Jabareen 2006; Eizenberg and Jabareen 2017). These strategies include: integration with public transport; the diversity or inclusiveness of urban landscapes; renewal and use of urban spaces; the compactness of the city; and optimisation of energy production and consumption.

A sustainable real estate sector can improve working conditions and provide employment opportunities in both developed and developing countries. With some estimates (World Economic Forum 2016) predicting that the sector will grow by 85% by 2030, this must have a positive impact on the labour market through direct job creation, both in new construction and modernisation. The labour market will also be positively impacted by the energy and civil infrastructure sectors, sustainable components and the operation and maintenance of energy efficiency programmes. This in turn will result in indirect employment benefits, such as recycling and waste management activities. Studies have argued that the sustainable real estate sector can generate more job opportunities than it replaces in the traditional energy supply industry (UNEP 2011).

Sustainable projects such as those described above, we argue, should be the key ambition of our communities, since cities are the intersection of economy, natural resource management, community and technology. The fact that developed countries are achieving better results than emerging countries in sustainable building development makes it essential to facilitate an international exchange of experience and knowledge.

2.7.1 Sustainable Activity in the Real Estate Industry

Socially responsible and sustainable approaches are increasingly being adopted by construction-sector companies with regard to the planning, construction and demolition of real estate projects (Bernstein et al. 2013). Owing to the decreasing cost of technology, increasing demand, and the proliferation of incentives for sustainable projects, the construction sector has more encouragement than ever to deliver green, sustainable and environmentally friendly buildings (Arbor 2005; Bernstein et al. 2013). A Deloitte report (2014) on commercial buildings shows that the implementation of ESG considerations in existing buildings can lead to high rates of performance relative to traditional practices. Barriers to the adoption of ESG principles within the real estate sector that have been reported include higher initial costs, lack of political support and lack of trained professionals (Bernstein et al. 2013).

The common pattern today in the sustainable construction sector is that it is driven by demand and commercial factors such as market transformation and local competition, and that investment decisions are influenced by high construction values and lower operating costs (Arbor 2005; Bernstein et al. 2013; World Economic Forum 2016). In addition, according to Bernstein et al. (2013), the increased focus on ESG aspects in many companies is due to a growing recognition of the value of brand opportunities. Those authors inform us that the most widely used sustainable products in the real estate sector are electrical, plumbing and waste-related, and that the most commonly used renewable energy source is solar (Bernstein et al. 2013).

The property valuation tools that are being used, although government-managed, are mostly voluntary third-party and privately operated (Ding 2008). The choice of such tools is certainly influenced by geographical location, while the economic-financial aspects may be a factor for investors as they are often not considered[6] (Ding 2008; Hens 2012). However, notwithstanding the risks for investors associated with valuation, the appropriate input from professionals will facilitate their use in chosen projects (DLA Piper 2014).

A study of the construction sector by Morrow et al. (2015) highlights the need to take not only environmental issues into account but also the productivity, health and well-being of the occupants. While it can be argued that existing assessment tools are already playing a key role in improving the ESG aspects in numerous projects, many scholars are of the opinion that a more comprehensive approach should be taken to guide design, increase environmental awareness, and impose more structure on the analysis of activities' environmental impacts (Ding 2008; Cucuzzella 2009, 2011; Fadaei et al. 2015). For these scholars, new techniques need to be adopted to evaluate buildings and real estate projects to take into account a wider range of considerations and to integrate ESG factors at the strategic decision-making level.

As we have already pointed out, recent trends have led to increased interest in sustainability in the real estate sector (Putnik 2009; Qian et al. 2015). But to properly integrate ESG aspects within the real estate sector will require a recognition of the complexity of these issues (Fisher 2008; Fry 2009). According to Fry (2009), for example, the achievement of

[6] In many cases, these tools provide multi-criteria matrices for evaluation (Ding 2008).

sustainability in the real estate sector requires nothing less than a reconceptualisation of the design profession. Other studies argue that process changes must also be considered, and that the design profession can serve as a guiding force for the creation of sustainable real estate (Walker 2006; Nelson and Stolterman 2012).

Although real estate is a risky sector, a rethink that incorporates the integration of ESG aspects can offer unique advantages. Such aspects must cover the full spectrum of the social, environmental, economic and ethical spheres (Ehrenfeld 2009) and also take into account changes in position and context, both temporal and geographical (Wilbanks 2007). To approach ESG seriously, it is necessary first to agree on a comprehensive system of sustainability principles (World Economic Forum 2016), which should include factors that affect the decision-making process and allow the optimisation and institutionalisation of all their constituent components (Ding 2008; Putnik 2009). ESG criteria should be integrated into all activities of the real estate life cycle, including community development and investment (Lawrence 2015). Each project must strengthen and connect those aspects that are to be developed on an individual, economy or society level with other areas to be supported such as the environment and the community (Robert et al. 2005).

References

Agarwal V, Narayan N (2004) Risk and portfolio decisions involving hedge funds. Rev Financ Stud 17(1):63–98. https://doi.org/10.1093/rfs/hhg044

AIMA (2018) From niche to mainstream: responsible investment and hedge funds. Alternative Investment Management Association. https://www.aima.org/article/pressrelease-responsible-investment-by-hedge-funds.html

Amaeshi K, Grayson D (2008) The challenges of mainstreaming Environmental, Social and Governance (ESG) issues in investment decisions: a mini-survey of practitioners' reports. http://citeseerx.ist.psu.edu/viewdoc/download?doi=10.1.1.466.1773&rep=rep1&type=pdf

Amin G, Kat H (2003) Hedge fund performance 1990–2000: do the money machines really add value? J Financial Quant Anal 38(2):251–274. https://doi.org/10.2307/4126750

Anson MJP, Fabozzi FJ, Jones FJ (2010) The handbook of traditional and alternative investment vehicles: investment characteristics and strategies. Wiley, Hoboken. https://doi.org/10.1002/9781118258248

Arbor A (2005) Building green for the future: case studies of sustainable development in Michigan. Handbook—Michigan University, Michigan

Baker HK, Filbeck G (2013) Alternative investments: instruments, performance, benchmarks and strategies. New York, Wiley. https://doi.org/10.1002/9781118656501.ch1

Bernstein HM, Russo MA, Fitch E, Laquidara-Carr D (2013) World green building trends—business benefits driving new and retrofit market opportunities in over 60 countries. McGraw-Hill Construction Research & Analytics, Bedford, MA. http://www.worldgbc.org/files/8613/6295/6420/World_Green_Building_Trends_SmartMarket_Report_2013.pdf

Bessler W, Wolff D (2015). Do commodities add value in multi-asset portfolios? An out-of-sample analysis for different investment strategies. J Bank Financ 60: 1–20. https://doi.org/10.1016/j.jbankfin.2015.06.021

Blanc D, Goldet B, Hobeika S (2009) Where do private equity businesses stand on the integration of ESG issues? Working Paper, Novethic SRI Research Center

Brandon P, Lombardi P (2010) Evaluating sustainable development in the built environment, 2nd ed. Wiley-Blackwell, Chichester, West Sussex

Brandstetter L, Lehner OM (2015) Opening the market for impact investments: the need for adapted portfolio tools. Entrepreneurship Res J 5(2):87–107. https://doi.org/10.1515/erj-2015-0003

Bukart K (2009) How do you define the 'green' economy? MNN—Mother Nature Network

Cambridge Associates (2015) Risks and opportunities from the changing climate: Playbook for the truly long-term investor. https://www.cambridgeassociates.com/research/risksand-opportunities-from-the-changing-climate-playbook-for-the-truly-long-term-investor

Cappucci M (2018) The ESG integration paradox. J Appl Corp Finan 30(2):22–28. https://doi.org/10.1111/jacf.12296

Chen P, Baierl GT, Kaplan PD (2002) Venture capital and its role in strategic asset allocation. J Portfolio Manage 28(2):83–89. https://doi.org/10.1002/9781119205401.ch15

Chen H-C, Ho K-Y, Lu C, Wu C-H (2005) Real estate investment trusts: an asset allocation perspective. J Portfolio Manage 31(5):46–55. https://doi.org/10.3905/jpm.2005.593887

Clark GL, Feiner A, Viehs M (2014) From the stockholder to the stakeholder: Oxford University and Arabesque partners study. Oxford University Press, Oxford. http://papers.ssrn.com/sol3/papers.cfm?abstract_id52508281

Crifo P, Forget VD (2012) Think global, invest responsible: why the private equity industry goes green. J Bus Ethics 116:21–48. https://doi.org/10.1007/s10551-012-1443-y

Cucuzzella C (2009) The limits of current evaluation methods in a context of sustainable design: prudence as a new framework. IJDE Int J Des Eng 2(3):243–261. https://doi.org/10.1504/ijde.2009.030174

Cucuzzella C (2011) Why is fourth generation evaluation essential for sustainable design? Des Princ Pract 5(1):239–251. https://doi.org/10.18848/1833-1874/cgp/v05i01/38014
Deloitte (2014) Breakthrough for sustainability in commercial real estate. https://www2.deloitte.com/us/en/pages/real-estate/articles/sustainability-in-commercial-real-estate.html
Ding GKC (2008) Sustainable construction—the role of environmental assessment tools. J Environ Manage 86(3):451–464. https://doi.org/10.1016/j.jenvman.2006.12.025
Dixon, D. (2009). Bringing ESG considerations into the investment process. Money Manage 23(13):12–12. https://www.moneymanagement.com.au/features/editorial/bringing-esg-considerations-investment-process. Accessed 24 Apr 2009
DLA Piper (2014) Towards a greener future—DLA Piper's market report on sustainable real estate. https://www.dlapiper.com/it/italy/insights/publications/2014/03/towards-a-greener-future/. Accessed 10 Mar 2014.
Ehrenfeld JR (2009) Sustainability by design: a subversive strategy for transforming our consumer culture. Yale University Press, New Haven
Eizenberg E, Jabareen Y (2017) Social sustainability: a new conceptual framework. Multidisciplinary Digital Publishing Institute. Sustainability 9(1):68. https://doi.org/10.3390/su9010068
Ennis RM, Sebastian MD (2005) Asset allocation with private equity. J Private Equity 8(3):81–87. https://doi.org/10.3905/jpe.2005.516861
Fadaei S, Iulo LD, Yoshida J (2015) Architecture: a missing piece in real-estate studies of sustainable houses. Procedia Eng 118:813–818. https://doi.org/10.1016/j.proeng.2015.08.518
Fink L (2018) Larry Fink's annual letter to CEOS: a sense of purpose. Blackrock. https://www.blackrock.com/corporate/investor-relations/larry-fink-ceo-letter
Fisher T (2008) Architectural design and ethics: tools for survival. Elsevier and Architectural Press, Amsterdam and Boston and London.
Friede G, Busch T, Bassen A (2015) ESG and financial performance: aggregated evidence from more than 2000 empirical studies. J Sustain Finance Invest 5(4):210–233. https://doi.org/10.1080/20430795.2015.1118917
Fry T (2009) Design futuring: sustainability, ethics, and new practice. Berg, Oxford, NY
Fulton M, Kahn BM, Sharples C, (2012) Sustainable investing: establishing long-term value and performance. Deutsche Bank. https://papers.ssrn.com/sol3/papers.cfm?abstract_id=2222740
Gond J-P, O'Sullivan N, Slager R, Homanen M, Viehs M, Mosony S (2018) How ESG engagement creates value for investors and companies. United Nations, Principles for Responsible Investing. https://www.unpri.org/aca

demic-research/how-esgengagement-creates-value-for-investors-and-companies/3054.article
Harji K, Hebb T (2010) Investing for impact: issues and opportunities for social finance in Canada. Carleton Centre for Community Innovation, Ottawa
Harji K, Jackson ET (2012) Accelerating impact: achievements, challenges and what's next in building the impact investing industry. The Rockefeller Foundation, New York, NY. https://www.rockefellerfoundation.org/report/accelerating-impact-achievements/
Harji K, Reynolds J, Best H, Jeyaloganathan M (2014) State of the nation—impact investing in Canada. MaRS, Toronto, pp 1–94. https://doi.org/10.13140/rg.2.2.36552.11526
Hedge Funds Review Editorial (2013) Hedge funds face dilemma of implementing socially responsible investing. Hedge Funds Rev. Accessed 20 Mar 2013
Hens H (2012) Passive houses: what may happen when energy efficiency becomes the only paradigm? ASHRAE Trans 118(1):1077–1085
Hua T (2011) ESG gains wider acceptance. Pensions & Investments. Accessed 24 Jan 2011
Intergovernmental Panel on Climate Change (IPCC) (2007) Climate change 2007: impacts, adaptation and vulnerability. Cambridge University Press, Cambridge. https://www.ipcc.ch/site/assets/uploads/2018/03/ar4-wg2-intro.pdf
Jabareen YR (2006) Sustainable urban forms their typologies, models, and concepts. J Plan Educ Res 26(1):38–52. https://doi.org/10.1177/0739456x05285119
Jackson ET (2013) Interrogating the theory of change: evaluating impact investing where it matters most. J Sustain Finance Invest 3:95–110. https://doi.org/10.1080/20430795.2013.776257
Jackwerth J, Slavutskaya A (2016) The total benefit of alternative assets to pension fund portfolios. J Financ Mark 31:25–42. https://doi.org/10.1016/j.finmar.2016.06.002
Jondeau E, Rockinger M (2006) Optimal portfolio allocation under higher moments. Eur Financ Manage 12(1):29–55. https://doi.org/10.1111/j.1354-7798.2006.00309.x
J.P. Morgan Social Finance (2014) The impact investor survey a spotlight on the market. https://thegiin.org/assets/documents/pub/2014MarketSpotlight.PDF. Accessed 1 May 2014
Lawrence M (2015) Reducing the environmental impact of construction by using renewable materials. J Renew Mater 3(3):163–174. https://doi.org/10.7569/jrm.2015.634105

Lee S, Stevenson S (2005) The case for REITs in the mixed-asset portfolio in the short and long run. J Real Estate Portfolio Manage 11(1):55–80. https://doi.org/10.1080/10835547.2005.12089711
Lindsay M (2013) Hedge funds need ethical guidelines. Hedge Funds Rev. Accessed 13 Feb 2013
LLNL (2014) Energy flow charts. https://flowcharts.llnl.gov/
Loew T, Clausen J, Hall M, Loft L, Braun S (2009) Case studies on CSR and innovation: company cases from Germany and the USA. Research report by future- verantwortungunternehmen, Institute4Sustainability, and Borderstep Institut für Innovation und Nachhaltigkeit. http://www.4sustainability.de/fileadmin/redakteur/bilder/Downloads/Publikationen/englische_PublikationenLoew_Clausen_etal-Case_Studies_on_CSR_and_Innovation.pdf
Lohr S (2011) First, make money. Also, do good. The New York Times. https://www.nytimes.com/2011/08/14/business/shared-value-gains-in-corporate-responsibility-efforts.html. Accessed 13 Aug 2011
Matisoff D, Noonan D, Flowers M (2016) Policy monitor—green buildings: economics and policies. Rev Environ Econ Policy 10(2):329–346. https://doi.org/10.1093/reep/rew009
McIntyre O (2015) Development banking ESG policies and the normativisation of good governance standards: development banks as agents of global administrative law. In Wendt K (ed) Responsible investment banking: risk management frameworks, sustainable financial innovation and softlaw standards. Cham, Switzerland, Springer International Publishing, pp 143–155. https://doi.org/10.1007/978-3-319-10311-2
Metrick A, Yasuda A (2010) The economics of private equity funds. Rev Financ Stud 23(6):2303–2341. https://doi.org/10.1093/rfs/hhq020
Michelini L, Fiorentino D (2012) New business models for creating shared value. Soc Responsib J 8(4):561–577. https://doi.org/10.1108/17471111211272129
Miles E (2010) Environmental, Social and Corporate Governance (ESG) and the UN PRI: Practical implementation: Environmental Resource Management (ERM).
Moore ML, Westley FR, Brodhead T (2012) Social finance intermediaries and social innovation. J Soc Entrepreneurship 3(2):184–205. https://doi.org/10.1080/19420676.2012.726020.
Morrow D, Read-Brown A, O'Sullivan N, Garz H (2015) Real estate: two steps forward, one step back. Sector Report. Sustainalytics. Thematic Research, September
Nelson HG, Stolterman E (2012) The design way: intentional change in an unpredictable world. The design way, vol 9, 2nd ed. MIT Press, Cambridge. https://doi.org/10.7551/mitpress/9188.001.0001

New Philanthropic Capital (2015) NPC'S manifesto 2015: a vision for change. https://www.thinknpc.org/publications/npc-manifesto-2015/. Accessed 9 Feb 2015

Nicklin S (2012) The power of advice in the UK sustainable and impact investment market. Bridges Ventures, London. https://www.bridgesfundmanagement.com/wp-content/uploads/2017/08/Bridges-2012-Impact-Report.pdf

O'Donohoe N, Leijonhufvud C, Saltuk Y (2010) Impact Investments: an emerging asset class. J.P. Morgan. http://www.rockefellerfoundation.org/uploads/files/2b053b2b-8feb-46ea-adbd-f89068d59785-impact.pdf. Accessed 29 Nov 2010

Ofori G (1992) The environment: the fourth construction project objective? Constr Manage Econ 10(5):369–395. https://doi.org/10.1080/01446199200000037

Porter ME, Kramer MR (2006) Strategy and society: the link between competitive advantage and corporate social responsibility. Harv Bus Rev 84(2):78–92. https://doi.org/10.1108/sd.2007.05623ead.006

PRA (2018) Transition in thinking: the impact of climate change on the UK banking sector. Bank of England, Prudential Regulatory Authority. https://www.bankofengland.co.uk/prudential-regulation/publication/2018/transition-in-thinking-the-impact-of-climate-change-on-the-uk-banking-sector

Putnik GD (2009) Complexity framework for sustainability: an analysis of five papers. Learn Organ 16(3):261–270. https://doi.org/10.1108/09696470910949971

Pwc, & Waterman (2010) Responsible investment: a guide for private equity and venture capital firms. BVCA, London, UK. https://www.bvca.co.uk/Portals/0/library/documents/BVCA_Responsible_Investment_Guide_2012.pdf

Qian QK, Chan EHW, Visscher H, Lehmann S (2015) Modeling the Green Building (GB) investment decisions of developers and end-users with Transaction Costs (TCs) considerations. J Clean Prod 109:315–325. https://doi.org/10.1016/j.jclepro.2015.04.066

Rice M, DiMeo R, Porter M (2012) Nonprofit asset management: effective investment strategies and oversight. Wiley, Hoboken, NJ. https://doi.org/10.1177/0899764013477378

Robert KW, Parris TM, Leiserowitz AA (2005) What is sustainable development? Goals, indicators, values, and practice. Environ Sci Policy Sustain Dev 47(3):8–21. https://doi.org/10.1080/00139157.2005.10524444

Rockefeller Foundation (2007) Bringing scale to the impact investing industry. https://www.rockefellerfoundation.org/blog/bringing-scale-impact-investing-industry/

Röhrbein N (2012) Hedge funds largely ignorant of ESG principles. Investments & Pensions Europe. Accessed 23 July 2012

Sa-Aadu J, Shilling J, Tiwan A (2010) On the portfolio properties of real estate in good times and bad times. Real Estate Econ 38(3):529–565. https://doi.org/10.1111/j.1540-6229.2010.00276.x

Saltuk Y, El Idrissi A, Bouri A, Mudaliar A, Schiff H (2014) Spotlight on the market—the impact investor survey. JPMorgan, London. https://thegiin.org/research/publication/spotlight-on-the-market-the-impact-investor-survey

Sassen R, Hinze AK, Hardeck I (2016) Impact of ESG factors on firm risk in Europe. J Bus Econ 86:867–904. https://doi.org/10.1007/s11573-016-0819-3

Schell JM (2020) Private equity funds: business structure and operations. Law Journal Press, New York, NY

Sparkes R (2002) Socially responsible investment: a global revolution. Wiley, s.l. https://doi.org/10.1002/9780470404324.hof002014

Staub-Bisang M (2012) Sustainable investing for institutional investors: risk, regulations and strategies. Wiley, Singapore. https://doi.org/10.1002/9781119199137

Steffensen H (2006) Environmental, social, & governance issues: materiality and relevance to investors

Teo MMM, Loosemore M (2001) A theory of waste behaviour in the construction industry. Constr Manage Econ 19(7):741–751. https://doi.org/10.1080/01446190110067037

The Impact (2016) Real assets and impact investing: a primer for families. https://arthaimpact.com/wp-content/uploads/2019/09/20094678-def5-4874-9f72-5ea646472c72_TheImpact_RealAssetInvestments_2016.pdf

Ulrich RS (1984) View through a window may influence recovery from surgery. Science 224(4647):420–421. https://doi.org/10.1126/science.6143402

United Nations Environment Programme (UNEP) (2011) Buildings: investing in energy and resource efficiency: towards a green economy. https://www.unep.org/greeneconomy/resources/green-economy-report

United Nations Population Fund (UNFPA) (2017) Urbanization overview. http://www.unfpa.org/urbanization

U.S. Green Building Council & Booz Allen Hamilton (2009) Green jobs study. U.S. Green Building Council, Washington, DC

Walker S (2006) Sustainable by design: exploration in theory and practice. Earthscan Publications, London. www.earthscan.co.uk

Wendt K (2015) Responsible investment banking. Risk management frameworks, sustainable financial innovation, and softlaw standards. Springer, Cham. https://doi.org/10.1007/978-3-319-10311-2

Wendt K (2018) Positive impact investing: a sustainable bridge between strategy, innovation change and learning. Springer International Publishing. 443.webvpn.fjmu.edu.cn/10.1007/978-3-319-10118-7_1

Wilbanks TJ (2007) Scale and sustainability. Clim Policy 7(4):278–287. https://doi.org/10.1080/14693062.2007.9685656

Wilson K (2014) New investment approaches for addressing social and economic challenges. In: OECD Science, Technology and Industry Policy Papers No. 15. https://doi.org/10.1787/5jz2bz8g00jj-en

Wood D, Hoff B (2007) Handbook on responsible investment across asset classes. Boston College Center for Corporate Citizenship, Boston, MA

World Bank (2016) Poverty overview. http://www.worldbank.org/en/topic/poverty/overview

World Economic Forum (2013) From the margins to the mainstream: assessment of the impact investment sector and opportunities to engage mainstream investors. World Economic Forum, New York. http://bit.ly/2jrQqIc. Accessed 11 Sept 2013

World Economic Forum (2016) Environmental sustainability principles for the real estate industry. Cologny, Geneva. https://www.weforum.org/agenda/2016/05/the-world-is-about-tosee-an-unprecendented-demographic-shift?utm_content=buffer8e511&utm_medium=social&utm_source=facebook.com&utm_campaign=buffer. Accessed 29 Jan 2016

Yau JK, Schneeweis T, Robinson TR, Weiss LR (2007) Alternative investments portfolio management. In: Maginn JL, Tuttle DL, McLeavey DW, Pinto JE (eds) Managing investment portfolios—a dynamic process, 3rd ed. Wiley, Hoboken, NJ, pp 477–578. https://doi.org/10.1604/9780470080146

CHAPTER 3

Socially Responsible Banking: Towards a New Firm–Bank Relationship

Francesco Gangi, Nicola Varrone, and Lucia Michela Daniele

3.1 A Schumpeterian View: The Social Role of Bank

In this section, we intend to answer a specific question: namely, whether a bank has a social function and, if so, how can this claim be justified?

A first and fundamental perspective is based on a Schumpeterian vision of the bank's role as an economic institution. Schumpeter was among the first to recognise a bank's crucial role in society, thanks to its function of supporting entrepreneurial initiative and, therefore, economic development. In acknowledging this role, Schumpeter makes reference to some fundamental aspects of his research, namely: the function of the entrepreneur; the interruption of the circular flow through innovation; and the development of the capitalist system. According to Schumpeter, bank credit acts as an injunction to the economic system to subordinate itself to the aims of the entrepreneur, entrusting him with resources (goods and productive forces) that he can use to further his initiative. Only in this way, says Schumpeter, does economic development take place; only in this way do we get to see the new "equilibrium" which comes about by doing the same things differently or by doing new things that interrupt the circular flow. In supporting the entrepreneur, the bank

F. Gangi et al., *The Evolution of Sustainable Investments and Finance*,
https://doi.org/10.1007/978-3-030-70350-9_3

fulfils a function that is the basis of what Schumpeter calls modern credit building.

Schumpeterian analysis is both forward-looking and innovative and transcends the long-standing debate about the mere creation of money. The basic concept on which Schumpeter builds his reasoning is the role of the bank in expanding the entrepreneur's purchasing power. The entrepreneur may be fortunate enough to be endowed with his own assets, but this is not a necessary prerequisite for an individual's entrepreneurial vocation. In case of insufficient purchasing power, the creation of means of payment is a fundamental innovation in the economic system which allows the entrepreneur to acquire goods produced by others, who are not willing to hand them over without receiving the payment of their selling price, nor are they willing to bear all the risk and wait for the entrepreneur to make a profit. The bank therefore fulfils this basic function of supporting entrepreneurial capacity, making available to the main engine of economic development—the entrepreneur—those resources that would otherwise not be available and which would end up delaying or entirely preventing the interruption of the circular flow. With no interruption of the circular flow, there would be no economic development.

The bank is, therefore, a source of innovation because it implies a management of the guarantees which otherwise would be entrusted exclusively to the commercial relationship between the parties. It is a source of innovation because it implies a different kind of risk management—a risk that would otherwise fall only on those who assign goods and services to credit. It is a source of innovation because, in addition to so-called normal credit, i.e. credit granted for goods previously delivered, it also grants "abnormal" credit, i.e. loans for future services. Schumpeter asserts that credit is essential for industrial development. The entrepreneur needs credit, understood as a temporary assignment of purchasing power, in order to be able to produce, to introduce new production combinations, and/or to become an entrepreneur in the first place. Schumpeter emphasises very clearly that any granting of credit for the purpose of innovation is, by definition, the granting of credit to the entrepreneur and constitutes an indispensable element of economic development.

Therefore, according to a Schumpeterian perspective, the first dimension of a bank's social function is the role it plays in "guaranteeing" the resources necessary for entrepreneurial initiative, the introduction of innovations and, therefore, the economic development of a country and its communities. This important interpretation leads us to the fundamentals of how the relationship between bank and company works, and for this we turn to the most recent literature. Wu and Shen (2013), for example, refer to Schumpeter in acknowledging that a properly functioning banking system boosts technological innovation by supporting those entrepreneurs with the best chance of success in introducing new products and/or new production processes. This analysis also works in the opposite direction, i.e. with regard to the negative impact that a credit restriction, or credit crunch, can have on economic development. Wu and Shen (2013) explain that the contraction in credit supply was one of the main causes of the US recession in 1991. Years earlier, King and Levine (1993) recognised that a healthy banking system is essential in sustaining prosperity. This is confirmed by Rajan and Zingales (2003), who claim there is ample evidence that a developed financial system and a strong economy go hand in hand, and that the development of banks encourages competition, reductions in interest rates, an increase in the rate of new business startups, and a greater availability of resources for financing economic growth. The two authors recognise that finance lubricates the process of economic growth and refer to Schumpeter (1934), according to whom modern industry would not have been born without finance. Without banks, the individual would have been dependent on inherited assets, whereas, thanks to banks and the financial system, talent can "ride" on its debts to success. Ultimately, the bank has an economic democratisation role: providing purchasing power to those who have none due to their origins, despite being endowed with entrepreneurial skills and vocation. As lubricants of the complex gears of economic development, banks fulfil an essential social function. According to de Bruin (2015), the counter-evidence is found in government interventions following the financial crises. In the US, after the speculative sub-prime mortgage bubble burst, American banks were bailed out with an effective injection of resources amounting to $7700 billion at rates close to zero by the Federal Reserve (*Il Sole 24 Ore*, 2 December 2011)—an intervention equal to about 50% of US GDP and more than ten times the value of the Troubled Asset Relief Program (TARP) launched by the Bush government and subsequently managed by the Treasury under the Obama presidency. The term "too big to fail" speaks to the indispensable function of banks, in particular those that the EBA (European Bank Authority) considers systemic banks because the welfare of entire economic systems may depend on their good health. The failure of Lehman Brothers and its dramatic effects is an experience that needs to be remembered.

Returning to our first point, a bank has the social function of bridging the purchasing power deficit for entrepreneurs. In Fisher's separation theorem (1930), banks play a fundamental role in decoupling investment decisions from financing decisions and, therefore, separating investment decisions from consumption decisions. Moreover, credit that nurtures talent makes an essential economic contribution that should never be underestimated. The intermediary function must be safeguarded because it reduces the gap between the richest and the poorest. It is an indispensable function, crucial in ensuring the efficient operation of the "social elevator".

A bank performs other, more operational, economic functions which are critical for social development (de Bruin 2015). Firstly, it implements a process of maturity transformation: savers generally deposit small amounts of resources for short periods of time, while borrowers may need more resources and for longer periods. Secondly, through this maturity transformation, banks also perform a risk management function. Instead of lending depositors' money to one particular borrower, they lend to a wide range of borrowers, implicitly diversifying lending and risk. A final function of banks is to contain transaction costs, also understood as research and knowledge costs. By virtue of banks, depositors do not have to divert resources into searching for borrowers and/or learning how to enter into contracts with them. In sum, the containment of transaction costs, risk diversification, and maturity transformation are the sub-functions that a bank performs while fulfilling its fundamental social function of contributing to a community's economic development and prosperity.

However, over time, banks' role as sources of economic development has changed. For one thing, the function of time transformation of maturities has been called into question at an institutional level, with reference, for example, to how the Germanic model has proved insecure for savers, something that had already become clear many years ago (Einaudi 1921). In other words, the logic of financing industrial development with long-term interventions must not compromise the nature of short-term deposit investments. This requires specialised financial institutions to be developed. Furthermore, over the years we have witnessed a quantitative reduction in banking interventions in favour of companies, which requires a new and more innovative conception of the banks' contribution to economic development.

3.1.1 *The Evolution of Bank Interventions in "Bank-Centred" Systems: The Italian Case*

Over the years, commercial banks' exclusive role of providing financial resources for the growth of economic systems has changed. This also seems to be increasingly evident among the so-called bank-centred economic systems that rely on banking interventions as a primary source of company financing.

To understand the direction of this evolution, let us use Italy as a reference point: an economic system that has always been based on the central role of banks (countries with similar characteristics are France, Germany and Spain). Based on GDP, Italy is the third largest country in the EU and the second largest country in terms of value of manufacturing output. Hence, Italy represents a very important economic entity in the international context, and especially so in the European one. Italy is the archetype of an economic and financial system deeply linked to the role of banks. There are several reasons for this dependency. For many years after the Second World War, the most important banks in Italy, what are known today as systemic banks, have been recognised as banks of national interest, with a strong influence of the public "hand" (the State and Government). Moreover, the stock market has always played a relatively marginal role in the Italian economy. More recently, however, its significance has grown, reaching 30% of GDP in 2019 (Consob, Bollettino Statistico 2020). To compare this with an economic system more aligned to the development of financial markets, the capitalisation of the US stock market reached 155% of GDP in 2018 (IMF), while, as of June 2020, the capitalisation of the top seven companies in the US market exceeded the GDP of Germany and Italy combined.

The other key factor which confirms the pre-eminence of banking in the Italian economic system is the low diffusion of venture capital and private equity. In fact, these markets accounted for just 0.25% of GDP in 2019 (Statista 2020), while the average across the main European countries is around 0.7%. Comparing these figures with that of the US, where the private equity market reached 5% of US GDP in 2018 (E&Y 2019), we see a significant discrepancy with both Italy and the rest of Europe.

The predominant role of banks in the Italian system is confirmed by these figures. However, a significant change is taking place with regard to the influence of banks relative to the rate of economic development, and we are even seeing this in Italy. Some figures will serve to confirm this evolution, allowing us to portray the role of the banking system in quantitative terms alone, relative to economic development.

First, from 2011 to 2019 (see Table 3.1), comparing Italian GDP with bank loans to companies, we notice a significant negative correlation

Table 3.1 Italian GDP and financial debts of firms

Italian GDP and financial debts (€/mln)

Year	*GDP (current prices)*	*Bank debts*	*Other. fin. debt*	*Bank leverage*	*Other fin. lev.*	*Banks/GDP*	*Other fin./GDP*
2011	1,648,756	894,257	376,101	59%	25%	54%	23%
2012	1,624,359	864,670	435,840	52%	26%	53%	27%
2013	1,612,751	813,914	455,050	52%	29%	50%	28%
2014	1,627,406	807,970	456,250	50%	28%	50%	28%
2015	1,655,355	793,495	485,420	47%	29%	48%	29%
2016	1,695,787	775,808	484,643	45%	28%	46%	29%
2017	1,736,593	726,612	520,317	40%	29%	42%	30%
2018	1,766,168	678,503	542,030	37%	29%	38%	31%
2019	1,787,700	631,239	592,297	34%	32%	35%	33%
Corr. index		−0.90***	0.85**	−0.90***	0.62*		−0.89**
Stand. dev.	65,488	84,788	63,225	8%	2%	7%	3%

(Variation)

Year	*Var. GDP curr. price (2011 = 100)*	*Var. bank debts (2011 = 100)*	*Var. other fin. debt (2011 = 100)*
2011	nd	nd	nd
2012	−1%	−3%	16%
2013	−2%	−9%	21%
2014	−1%	−10%	21%
2015	0%	−11%	29%
2016	3%	−13%	29%
2017	5%	−19%	38%
2018	7%	−24%	44%
2019	8%	−29%	57%
Corr. index		−0.95***	0.95***
Stand. Dev.	4%	9%	14%

Sig. Level (*) 10%; (**) 5%; (***) 1%
Data Source ISTAT; Banca d'Italia

(−0.9), while, when comparing Italian GDP with other financial debts of companies, the correlation is significant and positive (0.85). Given the R values (>0.5), the correlation is in both cases very high and very intense. Second, if we compare the performance of Italian GDP with the level of bank leverage and the leverage linked to other debt instruments, the correlation remains significantly negative in the first case (−0.9) and significantly positive in the second (0.62). In both cases, the R values express a high and intense correlation. Third, if we compare the ratio between bank loans and GDP with the ratio between other financial loans and GDP, we note that the correlation is significant and negative (−0.089). The relationship is high and intense. Finally, if we compare, in a dynamic sense (2011 = 100), the variation of Italian GDP with the variation of bank loans and with the variation of other financial debts, we note that, in the first case, the correlation is significant and negative (−0.95), while, in the second case, the correlation is significant and positive (0.95). In these cases, too, the R index indicates a very high and very intense correlation.

This analysis confirms the relationship between the evolution of banking interventions compared to the economic growth trend in a "bank-centred" system using the case of Italy. This correlation has been negative over the last nine years, whereas the correlation between economic growth and the performance of other corporate loans (e.g. bonds) has been positive. This duality testifies to a "divorce" between banking interventions and the quantitative development of an economic system such as Italy's which, historically, has been centred on the role of banks.

Three considerations emerge from this. First, the role of the bank in relation to economic development is less significant from a quantitative point of view Second, increasing space has appeared for other financing instruments for companies. Third, the contribution that the banking system can make to general economic evolution should be understood not only in terms of quantitative growth but also in terms of qualitative development. In other words, attention must be turned to the qualitative contribution that the banking system can make, such as the impact on sustainable economic development.

3.2 Why Do Banks Comply with ESG Principles?

In the movie *It's a Wonderful Life* (Capra 1946), banker George Bailey is saved from an imminent bankruptcy on account of a scam by cynical banker Henry Potter, thanks to the collective effort of the citizens of Bedford Falls, whose way of life would have been much less prosperous without George Bailey and his bank. George Bailey embodies what we understand here as the social function of the bank. Nowadays, this function is more frequently addressed by the intentional application of the triple-bottom-line (TBL) principles (Elkington 1997) of corporate social responsibility (CSR), which leads to a new vision of what we call "socially responsible banking" (SRB). SRB has social, economic and ethical implications, which we will address in this section.

Once we have clarified why banks perform a social function in the economic development and the welfare of a community, there is a further, fundamental, dimension of a bank's social role that we need to look at, and this is its contribution to sustainable development. This concept of sustainability includes social and environmental sustainability, in compliance with TBL principles, which include profitability. The concepts of sustainability and CSR are closely interrelated and frequently overlap (Montiel 2008; Carvalho et al. 2014). Sustainability means safeguarding resources for future generations, which implies enhanced community well-being. Accordingly, in line with Renneboog et al. (2008, p. 1723), we define CSR as "corporate decisions fostering social, corporate governance, ethical and environmental issues". To measure bank social performance (BSP), therefore, we gauge the extent of its engagement in social, environmental and corporate governance issues.

There are two answers to the question of why and how banks can act in accordance with CSR principles. First, it is because this approach is consistent with their role as an engine for the quantitative and qualitative development of a given economic system. Second, it is because compliance with CSR principles can improve their performance and reliability. As we can demonstrate, these two answers are two sides of the same coin.

Starting with the first answer, we first note Scholtens's (2009) observation that socially responsible banking is becoming a well-established phenomenon in the financial services industry; banks are becoming increasingly involved in the financing of economic activities that aim to promote sustainable development. This trend is consistent with their role and function, which, as financial intermediaries, is much closer to a company's business model than that of the stock markets. In providing money to business initiatives, banks play an essential role in selecting the best projects and supporting their implementation.

Returning to Schumpeter's (1934) concept of financing talent, this can also be understood from a social responsibility perspective. A talented entrepreneur can be someone who promotes socially and environmentally sustainable business ideas and so a bank can play an important role in selecting such talent. Financially supporting such initiatives contributes to the dissemination of activities with social, ethical and environmental benefits. Indeed, this is a new way of understanding manifold support for economic development by banks.

The bank plays a crucial role in project screening, and it can assume that initiatives that are not sufficiently socially responsible may be penalised by being denied bank financing. Banks can therefore use credit to foster greater awareness of the importance of sustainable business models, creating a dual effect for the dissemination of CSR. On the one hand, banks apply CSR principles to their activities; on the other, promising socially and environmentally oriented companies and projects can be rewarded on the basis of their compliance with CSR. This dual effect, in which a diffusion of CSR culture leverages loans, represents a new contribution that a bank can make to economic development, if, by development, we also mean an improvement in the quality of the sustainability of products and production processes. In practice, then, CSR-sensitive banks can install additional requirements around the granting of credit anchored to the degree of compliance with CSR principles and the way in which an entrepreneur embeds social and environmental responsibility in the planning and management of his/her business.

This is a modern understanding of the functioning of banks that is not at odds with their intrinsic nature as financial intermediaries. On the contrary, as Wu and Shen (2013) point out, banks enjoy the use of a considerable amount of society's resources, so it is reasonable to expect them to give something back to the community at an equal, if not higher, level than other economic agents. It is high time to retire the outdated conception that sustainability is the concern only of manufacturing industries and, in particular, those with the highest environmental or social impact (e.g. energy, oil, mining, armaments). Banks are closer to their clients' business models than other categories of lender. They must therefore manage sustainability in the same way as other sectors, and their involvement in CSR initiatives can be both direct as well as indirect, through their lending channels (Weber 2012). As such, banks have the opportunity to be more actively involved in environmental protection and the reduction of environmental risks. A recent *Financial Times* article (Hook 2018) pointed out that the European institutions that supervise banking activities are showing a growing interest in understanding banks' exposure to climate risks and are looking at adapting the risk-assessment regulations related to environmental damage, which includes the use of stress tests on fossil fuel exploitation and CO_2 emissions. Specifically, the European Banking Authority (EBA) has confirmed it will develop a climate stress test and further explore the possibility of offering banks favourable prudential treatment for lending to green projects. The EBA's Action Plan on Sustainable Finance (APSF), launched in 2019, urged financial institutions to act immediately to incorporate ESG factors into their business strategies. Its latest plan aims to monitor climate risk by looking at strategy and risk management for member bodies, with an emphasis on metrics and bank disclosure. Consistent with the scope of APSF, the successful measurement of ESG risks by banks can form the basis of scenario analysis and an explorative tool to understand the exposures and the potential magnitude of ESG risks. This aim includes the objective of a dedicated climate change stress test.

The Bank of Italy's annual report (2020, p. 35) now includes a section on climate risks, with the Governor pointing out that "there is ample scientific evidence about the effects of climate change, which have important economic repercussions. According to some analyses, the spread and social costs of pandemics could be increased by environmental pollution associated with the use of fossil fuels; the current health emergency would therefore make the reduction of greenhouse gas emissions even more important."

At the institutional level, the increase in initiatives aimed at raising banks' awareness of the application of ESG criteria further confirms their social role in terms of their contribution to sustainable economic development. For example, as of August 2020, more than 180 banks have joined the United Nations Environment Programme Finance Initiative (UNEP FI). This programme includes the assessment of different areas of risk and performance related to the environment, including climate risk (e.g. environmental disclosure, portfolio carbon risk, climate performance), and natural capital risk and evaluation (e.g. development of environmental risk management, removing deforestation from commodity value chains,

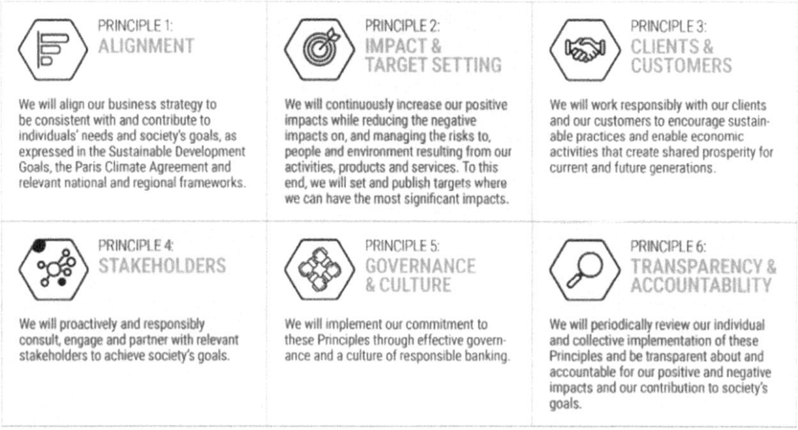

Fig. 3.1 The UNEP FI principles for responsible banking (*Source* www.unepfi.org)

and access of agricultural smallholders to finance). This UN programme confirms the crucial role of the banking system in promoting sustainable development. The UN has worked with programme members to define the Principles for Responsible Banking (PRB), "which aim to transform the banking industry to enable it to play a leading role in achieving society's goals as expressed in the Sustainable Development goals and the Paris Climate Agreement".[1] At the time of writing, 185 international banks had signed up to the PRB initiative. The principles include six areas of assessment and commitment for banks to consider in carrying out their lending function: alignment; impact and target setting; clients and customers; stakeholders; governance and culture; transparency and accountability (see Fig. 3.1).

Historically, the other important international initiative that speaks to the growing recognition of the social role of banks in sustainable development is the Equator Principles (EPs), to which ten international banks first signed up in 2003. Today, more than 100 banks from over 30 countries participate in the EPs, the general purpose of which is to provide "a risk management framework, adopted by financial institutions, for determining, assessing and managing environmental and social risk in projects

[1] https://www.unepfi.org/.

and is primarily intended to provide a minimum standard for due diligence and monitoring to support responsible risk decision-making".[2] The EPs, increasingly widely applied by banks, can claim to have brought increased attention to social standards and local communities, and they have boosted the development of socially and environmentally responsible management practices within the financial sector and the banking industry in particular. In fact, a bank, having signed up to the EPs, is called on to make thorough evaluations of the positive and negative externalities of all projects being financed.[3]

With regard to the EPs—although a similar point can be made with reference to the PRBs and UNEP FI in general—Scholtens and Dam (2007) identify three aspects, namely: the risks of "greenwashing", CSR over-compliance, and CSR as business-as-usual. The first point refers to the possibility of banks signing up to guidelines such as the PRBs or EPs only with the intention of improving their image but without applying them—or any standards of social responsibility—in practice. According to Wu and Shen (2013), banks that undertake greenwashing make no substantial changes to their business models and at the same time lose out on the potential benefits related to the implementation of CSR policies. In the second point—over-compliance—in contrast, banks incur the highest costs of EP or PRB guidelines, or similar schemes. In theory, these principles can restrict banks' markets and increase their opportunity cost of capital. Additional screening and monitoring for social and environmental objectives can incur additional operating costs, not only for the banks but also for the projects to be funded, which must therefore generate a higher financial performance to offset their higher CSR-related operating costs. However, where these incremental costs are offset or even outweighed by the benefits arising from the application of the CSR guidelines, at that point they may become the new business-as-usual (the third point).

Some pages ago we said there were two answers to the question of why and how banks can act in accordance with CSR principles. The arguments above have now guided us towards the second answer. It centres around the issue of what advantages or factors can make the benefits of CSR outweigh its costs, and such justification must include not only financial but also social and environmental considerations. As an overview of how

[2] https://equator-principles.com/.

[3] For an examination of the history of EPs, see Esty et al. (2005).

Table 3.2 Diffusion of voluntary principles for socially responsible banking

Diffusion	*Principles-schemes*	
	Equator principles	*UNEP-FI*
Adhesion of banks ($n°$)	108	185
Regions ($n°$)	7	5
Countries ($n°$)	>30	>50

Source Authors' calculation on data collected from Equator Principles and UNEP FI web sites

far such social and environmental considerations have been disseminated internationally, see the data on banks' adherence to voluntary principles in Table 3.2.

3.2.1 *ESG Principles in Banking as Drivers of Business Benefits*

The social role of banks has a number of facets. We have discussed the contribution that banks make to economic development, and to the spread of more sustainable business models. There is also a third aspect: the safeguarding of savings. As already highlighted (Wu and Shen 2013), banks have a moral as well as an economic duty to protect the interests of depositors who rely on banks' management to protect their resources. Banks use resources from the public and for this reason they are accountable for their actions to a large number of people. From the correct functioning of a bank derives the well-being of the community that has entrusted to it a share of its savings and which considers the bank the main financial intermediary in supporting its projects. All this is based on people's trust in the bank's reliability. Trust and reliability are therefore the pillars on which the relationship between a bank and its creditors is based. Thus, it is important to understand how CSR can become the lubricant of the multiple relationships between bank and customers, both depositors and entrepreneurs.

One CSR-related factor that can strengthen trust between a bank and its clients is reputation (Wu and Shen 2013; Forcadell and Aracil 2017). A bank's commitment to CSR contributes to an improved image in the eyes of various stakeholders. And the creation of long-term relationships of trust enables a bank to strengthen its ties with customers. Banks with stronger reputations are seen as more reliable and are preferred by those clients who are more sensitive to CSR issues. Such a preference has several economic implications. First, the most reputable banks can offer services with more advantageous economic conditions. In turn, customers will be willing to pay a higher price for those services when provided by a bank with the best reputation for reliability. With regard to motivational aspects, in looking at socially oriented clients who are highly interested in gauging a bank's commitment to CSR, Perez and Rodriguez del Bosque (2013) confirm that the greater a bank's commitment to social and environmental interests—directly and indirectly—the more likely a CSR-oriented client is to prefer it. Further, banks with a stronger reputation tend to be seen more favourably by companies too, who acquire a certification effect through the bank's reputation. Thus, for the same price paid, companies obtain a dual service, i.e. financing provision and certification of their projects. Being able to provide this dual offer can increase the bank's market power and business volume. Finally, reputation can lead to a virtuous cycle to the extent that the most reputable banks tend to be more selective in order to safeguard their image (Chemmanur and Fulghieri 1994), and this selectiveness is also manifested from a social and environmental sustainability point of view. Saunders and Allen (2002) note that reputation is crucial factor in a financial project and that banks must avoid the damage their reputation can suffer on account of improper behaviour by the companies they finance.

Reputational considerations, and the link with CSR, have leapt to prominence after the 2008 financial crisis, which uncovered the pathologies inherent in the dark side of finance. It led to a major loss of confidence in a banking system that was guilty of economically and socially irresponsible and opportunistic behaviour. The increased commitment to CSR has contributed to a recovery of the confidence that was lost and the trust in reliability that had been compromised, as events after 2008 have confirmed. Shen et al. (2016) highlight CSR's role in the recovery of trust after the scandals involving Goldman Sachs in 2010, Merrill Lynch in 2011 and Barclays in 2012. Committing to CSR should steer banks away from exploitation of their clients and towards gaining their trust. Quoting a study by Dombret and Kern (2003) on commercial banks, de Bruin (2015) recalls that before the 2008 financial crisis very few banks knew their clients well or understood their needs. The rehabilitation of this relationship, too, can be achieved through a greater commitment to CSR (Perez and del Bosque 2013). CSR presents a fundamental opportunity to learn about the social and environmental issues connected with the implementation of projects being funded (Esty et al. 2005; Scholtens and Dam 2007).

Based on the assumption that in the banking industry security is preferable to profit (Shen et al. 2016), a second factor linking CSR to bank reliability is risk management (Jo et al. 2015). The link between CSR and risk is twofold. On the bank side, a greater commitment to CSR can reduce the risks of boycott by clients, of environmental scandals, of actions by employees, and of pressure from NGOs (Scholtens and Dam 2007). On the client company side, a greater commitment to CSR reduces the social and environmental risk of its business model (Cai et al. 2016). Recent surveys (Gangi et al. 2019b, 2020) confirm that a greater commitment to CSR leads to a lower exposure to the risk of financial distress on both sides of the equation—companies and banks—which includes small and medium enterprises which are normally more exposed to such risks.

Commitment to CSR therefore activates a virtuous cycle between bank and company, both parties reducing risks and therefore increasing reliability. This synergistic aspect is also very important from an ethical point of view. From a Kantian perspective (1775–1780), one of the three ethical imperatives is prudence (the others are skill and morality). Prudence is defined by Kant as the ability to determine an end (happiness), together with the means to achieve it. That is, for Kant, prudence is dexterity in one's use of the means in pursuit of the universal end-goal of man: that is, happiness. If we are therefore to see banks as being great facilitators not only of economic growth but also of individual happiness (de Bruin 2015), the importance becomes even clearer of respecting the ethical imperative of prudence in order to guarantee a bank's health and at the same time safeguard the welfare of the community. Recent experience confirms that, while acknowledging the social function of the bank, the fulfilment of the prudence imperative cannot be taken for granted. And this depends on those who head up the financial institutions. Shortly before the outbreak of the "quiet" sub-prime mortgage bubble (Hong and Sraer 2011), some major US banks were doing anything but fulfilling their prudence obligations. Merrill Lynch managed around $1 trillion of assets with an equity of $30 billion; Bear Stearns had about $12 billion of equity compared to $388 billion of overnight market exposure (Financial Crisis Inquiry Commission Report 2011). In fact, the 2007–2008 financial crisis affected banks' business models in a variety of ways. Paulet et al. (2015) highlight the demarcation between conventional and so-called ethical banks. Conventional banks have demonstrated two different kinds of behaviour: some were able to anticipate changes by adopting a more responsible financial attitude; others merely changed their behaviour in accordance with the new regulatory framework. Ethical banks, on the other hand, did not significantly alter their behaviour in relation to the new financial environment. This variation in responses to the effects of the financial crisis leads Paulet et al. (2015) to identify the existence of different business models within the two types of banks. We will return to this at the end of the chapter.

A third factor linking CSR commitment to economic benefits is the brand recognition effect. Recently, the increasingly frequent advertising campaigns by banks such as Banca Intesa or BNL of the BNP Paribas Group have been based on the message of the bank's commitment to social responsibility. Banca Intesa's 2020 advertising campaign proclaims "the story of its ongoing commitment to sustainable and inclusive development. The Group works every day to create an economic system in which everyone can express his or her potential and acts as a social and cultural reference model". BNL's 2020 advertising campaign is about sustainability as a factor of social as well as economic development, with a particular focus on socially responsible funds. Differentiation through brand recognition is clearly an area in which banks see their approach to sustainability as playing a major role.

Other channels through which CSR can produce economic benefits for a bank have been identified by Shen et al. (2016). First is a better exploitation of the interest margin due to the increased reputation enjoyed by its clients. Clients may be willing to pay more for the certification effect or may charge less if they consider the bank a sufficiently reliable place for their deposits. Also, a socially responsible bank can curtail overheads by reducing advertising expenses or costs incurred because of conflicts with clients.

Finally, an increased CSR commitment can make the bank more attractive to the best human capital. The best-qualified people will always be inclined to work for the most reliable banks with the strongest reputation and not for the less reputable or less reliable ones. This advantage was even recognised by Friedman's (1970) notorious analysis of the effects of CSR. In addition, attracting and retaining the best human resources facilitates fulfilment of the ethical imperative of "skill", i.e. the ability to use certain means in pursuit of an end, namely: *the welfare and happiness of people*.

So, overall, the second answer to why banks adopt management principles consistent with social responsibility has valid and robust theoretical and empirical foundations. Demonstrating the positive link between bank social performance (BSP) and bank financial performance (BFP) is important because it confirms the virtuous cycle created in the relationship between banks and socially responsible companies. This link is crucial because it can make a decisive contribution to creating the necessary conditions for sustainable economic development which further strengthens the social role of the bank.

3.3 An Overview of ESG Criteria in the Banking Sector

Over recent decades, academics have been paying increasing attention to socially responsible initiatives, and if and how banks engage in them (e.g. Wu and Shen 2013; Wu et al. 2017). As Scholtens (2009) tells us, the by now numerous studies on CSR in the banking industry have made social responsibility in banking a "well-established concept". As Goyal and Joshi (2011) put it, socially responsible banking focuses on the satisfaction of needs in the real economy and in society, in doing so also taking into account social, cultural, ecological, and economic sustainability.

The banking sector is highly relevant in society for a number of reasons (Lentner et al. 2015). First, banks collect resources from society in the form of deposits and use those resources to finance economic activities by lending and through capital markets. Acting as financial intermediaries, banks reduce information asymmetries between market operators, favouring the allocation of resources. Costs may be associated with gathering and understanding information about possible investments but an individual saver may not have the required resources to evaluate companies and market conditions and to allocate capital. By making information costs economical and purveying technical financial knowledge, the banking industry improves the efficiency of capital allocation (Levine 2005; Greenwood and Jovanovic 1990). Second, banks perform a monitoring and governance function on the companies they finance, potentially influencing managers' decisions and improving the efficiency of firms' investments. Prior literature recognises that a well-functioning financial system, through the improvement of corporate governance mechanisms, may enhance firms' productivity and promote economic development (e.g. Bencivenga and Smith 1993). Banks also assume a crucial role in several other activities such as the reduction and management of risk, providing the vehicles for risk trading and diversification, or the organisation of payment systems and the pricing of financial assets (Greenbaum and Thakor 2007). On account of the important connections that these activities have with wider society, banking more than any other industry is called on to demonstrate that it is operating in line with societal expectations.

Moreover, the banking industry also has its own specific peculiarities and distinctive traits that make the social responsibility of banks worthy of special attention. For instance, it is a highly regulated industry in terms of capital adequacy, ownership restrictions, activity restrictions, external auditing requirements and provisioning requirements (Levine and Barth 2001). Also, banks are frequently referred to as knowledge-intensive organisations, employing as they do a large proportion of highly qualified staff (Drucker 1993; Saengchan 2008; Curado 2008). Moreover, due to the intangible nature and the high technical content of its products, the banking business may be characterised by higher information asymmetries than other sectors (de Andres and Vallelado 2008). All these characteristics add up to differentiate banking from other sectors. As a consequence of this, engagement in socially responsible initiatives may also look very different (Lentner et al. 2015). Prior literature already recognises that, with different industries facing different stakeholder expectations (Rowley and Berman 2000), CSR conceptualisation and operationalisation may also vary across industries (Decker 2004). With this background in mind, this section intends to provide an overview of bank social responsibility, discussing the areas of greatest relevance for each pillar of ESG performance.

With regard to the "E" for "environment" pillar, despite some authors arguing that it has a marginal relevance for banks, since they are less polluting compared to other industries (e.g. Esteban-Sanchez et al. 2017), prior studies have ascertained that the banking industry can play a central role in the field of environmental sustainability (e.g. Gangi et al. 2018). Similar to other industries, banks can have a direct impact on the environment through waste management, the consumption of energy, and suppliers' compliance with environmental standards. But, more importantly, banks also have an indirect impact, relative to the environmental and social records of the companies that they finance

(Thompson and Cowton 2004). As recognised by Calderon and Chong (2014), environmentally and socially responsible lending represents a fundamental component of bank social responsibility. The authors define socially responsible lending as a bank's commitment to lend only to firms that integrate environmental and social issues into their operations.

Prior literature suggests that an inclusion of environmental performance information in credit risk models can reduce information asymmetries (e.g. Goss and Roberts 2011; Costello and Wittenberg-Moerman 2011; Dhaliwal et al. 2011) and thus improve the quality of debts contracted. There is growing empirical evidence that banks have begun to incorporate environmental issues in creditworthiness assessments (e.g. Herbohn et al. 2019) and are showing a tendency to offer lower interest rates to firms with good environmental performance (e.g. Goss and Roberts 2011; Nandy and Lodh 2012; Jung et al. 2016; Du et al. 2017). This trend reflects the growing impact that the environmental dimension is having in terms of firm performance (e.g. Shin et al. 2018). Prior studies demonstrate that engagement in environmentally responsible practices may generate both tangible and intangible benefits (Kolk 2016; Montiel 2008; Oikonomou et al. 2012), potentially improving a firm's competitiveness (e.g. Porter and van der Linde 1995) and financial performance. As a logical extension, banks that integrate environmental issues within credit risk models may benefit from a higher quality of loans portfolio due to the presence of companies that are less risky and deliver better economic performance.

From an evolutionary point of view, Jeucken and Bouma (1999) identify four approaches that banks can adopt towards environmental responsibility. First is a defensive approach, wherein the bank takes no action to support environmental regulation or else action is delayed. Second is a preventive approach, which takes environmental issues into account only in internal practices. Third is an offensive approach, wherein banks integrate environmental issues both in internal practices and towards external stakeholders. Finally, a fourth approach implies that banks should not lend to firms whose products are environmentally harmful.

Like environmental responsibility, the "S" for "social" pillar also has specific relevance for the banking industry. In this context, community involvement represents a cornerstone of bank social responsibility. This, according to Brammer and Millington (2003), extends far beyond charitable initiatives to include numerous forms of monetary and non-monetary contributions to the local community. Esteban-Sanchez et al.

(2017) see community involvement as referring to the influence that the banking industry has on the community through philanthropy, investments, and compliance with standards of business ethics. In recent years, banks have introduced several initiatives that signal a growing engagement with society and local communities. BNP Paribas UK, for example, encourages its employees to dedicate one or more days per year to volunteering and has established volunteering goals for each department. Barclays is engaged in several initiatives, ranging from helping the jobless find work to supporting social enterprises. BBVA's integration scholarship programme aims to help children and young people to return to school and complete their education. Among other initiatives, we can include the European Banking Federation's (2008) financial support for art and culture, academic and research scholarship, support for drug prevention initiatives, and sponsorship of young entrepreneurs.

Numerous banks are now rolling out financial inclusion programmes. The World Bank (2020) explains that financial inclusion translates into giving people access to financial products that meet their needs. These initiatives typically encompass the provision of solutions and products for those categories of people that might be not fully integrated into society or have a restricted level of access, such as migrants, low-income families, temporary workers and micro-enterprises. Among the main services offered under financial inclusion are microfinance, migrant banking and financial education. Another sub-dimension of the social pillar is banks' relationships with their employees. In this context, prior literature (e.g. Esteban-Sanchez et al. 2017; Gangi et al. 2018) as well as ESG rating agencies (e.g. Asset4) include employment quality, employee health and safety, training and development, and diversity and equal opportunity. Several prior studies have shown that CSR activities targeted at employees have a positive effect on their engagement and motivation (e.g. Rupp et al. 2018). Other studies show that positive employee perceptions of an employer's CSR engagement enhance job satisfaction and organisational commitment (Asrar-ul-Haq et al. 2017). Déniz-Déniz and Saá-Pérez (2003) found Spanish savings banks' institutionalised employee CSR initiatives had a positive effect on profitability. More recently, focusing on a sample of European banks, Gangi et al. (2018) argue that a bank's engagement in socially responsible activities targeted at its employees will favour the sharing of CSR knowledge within the organisation, which is a prerequisite for the effective implementation of CSR targeted at external stakeholders.

Finally, with the regard to the "G" for "governance", socially responsible banks need corporate governance that responds to stakeholder demands and effectively integrates environmental and social issues in strategic decision-making. There is a recognition in the literature of the connection between corporate governance and engagement in socially and environmentally responsible activities (e.g. Jo and Harjoto 2011; Harjoto and Jo 2011), although the conceptualisation of the relationship between corporate governance and CSR has taken time to evolve. On the one hand, the traditional perspective, also known as the narrow view, sees corporate governance as a set of rules that defines how decisions are taken in the organisation, in order to act in the best interest of shareholders (Shleifer and Vishny 1997; Saravanamuthu 2004). This view of corporate governance may be theoretically linked to agency theory (Dalton et al. 2007), in which, as we have seen, managers may have interests that diverge from those of the shareholders (Jensen and Meckling 1976) and may operate to pursue their own interests (Barnea and Rubin 2010). Against this backdrop, corporate governance represents the system through which shareholders monitor management's decisions (Shleifer and Vishny 1997). On the other hand, the broader view argues that corporate governance represents a dimension of CSR, going beyond the framework of how power is exerted within a company (Hancock 2005; Jamali et al. 2008). This position, also known as conflict resolution hypothesis (e.g. Calton and Payne 2003; Windsor 2006; Harjoto and Jo 2011), states that effective corporate governance systems, alongside CSR engagement, mitigate the conflicts between managers and stakeholders. From this angle, good corporate governance systems, coupled with engagement in socially and environmentally responsible activities, allow stakeholders' interests to be taken into account in a more effective way.

3.4 ESG Investment Policy and Bank Lending

Institutional equity ownership has grown significantly in recent decades (Sethi 2005). Today, institutional shareholders play a major role in the equity market, investing in a number of large corporations from developed countries (Cox et al. 2008). This phenomenon is mirrored in the SRI segment, where institutional investors hold almost 70% of the market share (Eurosif 2018). Alongside this trend, the impact that the presence of socially responsible investment funds (SRFs) among shareholders has on target companies has attracted increasing attention from scholars (e.g. Jamali et al. 2008): several studies have investigated how a firm's engagement in CSR is associated with the ownership of SRFs (e.g. Ahmed et al. 2012). Since CSR can be considered a form of strategic investment (Porter and van der Linde 1995; McWilliams et al. 2006), it is likely that SRFs, like other institutional shareholders, are involved in decisions about whether or not to engage in socially responsible initiatives (Oh and Park 2015). Starting from this basic tenet, the literature has attempted to delve more deeply into how SRFs affect target companies' CSR engagement (e.g. Faller and zu Knyphausen-Aufseß 2018).

By their nature, SRFs allocate resources collected from unitholders to firms compliant with ESG standards, thereby affecting the availability of capital for target companies. Consequently, one of the most obvious reasons driving companies to engage in CSR is the desire to access a bigger supply of equity capital (Haigh and Hazelton 2004; Wahba 2010; El Ghoul et al. 2011; Gupta 2018). However, prior studies claim that SRFs are still not widely diffused and their market power is too limited compared to conventional institutional shareholders to have any significant influence on corporate strategic decision-making processes (Statman 2000; Scholtens 2009). According to Johnsen (2003), SRFs can have an impact as long as they are actively engaged in improving a firm's performance in the area of social responsibility, voting in shareholders' meetings or establishing a dialogue with management. In this regard, Dimson et al. (2015) noted that the effectiveness of institutional investors' activism on ESG issues depends on firms' capacity to implement changes and is more likely for firms with reputational concerns. Other authors suggest that a long-term orientation and institutional shareholders' greater ability to monitor a firm's performance can induce managers to invest in areas that are likely to increase firm value over the long term, such as social responsibility (e.g. Graves and Waddock 1994; Mahoney and Roberts 2007).

Moreover, since it is recognised that CSR engagement positively affects profitability (e.g. Barnett and Salomon 2012) and mitigates firms' risk (Cai et al. 2016), we might expect institutional shareholders to trigger the undertaking of socially responsible initiatives in order to enjoy the benefits of enhanced financial performance (Dam and Scholtens 2012). However, other authors recognise that CSR engagement requires expenditure, compensation for which will come only from uncertain and/or long-term revenue (Jia and Zhang 2013). Thus, it is likely that institutional shareholders support engagement in socially responsible initiatives but only to the extent to which the achieved financial benefits exceed the costs incurred by the allocation of capital in CSR activities (e.g. Barnett and Salomon 2006).

Against this backdrop, many studies have empirically investigated the effects that SRFs have on target companies' CSR orientation (Dyck et al. 2019). However, to date, the findings remain either inconclusive or contradictory. For example, focusing on a sample of Egyptian firms, Wahba (2010) finds that institutional ownership stimulates companies to adopt environmental management standards. Similarly, examining a sample of over 500 UK firms, Cox et al. (2004) find that institutional investments positively influence target firms to engage in socially responsible initiatives. Conversely, Muniandy and Barnes (2010) and Ahmed et al. (2012) find no significant relationship between CSR engagement and institutional ownership.

As well as institutional equity ownership, there are other financial instruments, such as bank lending, playing a key role in bringing ESG issues into economic activities (Scholtens 2009). As we have discussed, and as prior studies recognise (Simpson and Kohers 2002; Thompson and Cowton 2004), banks need to consider the social implications of their lending decisions. Their lending practices are usually based on credit score systems, which include quantitative and qualitative information useful in assessing a borrower's creditworthiness. While quantitative information will mainly be financial, qualitative information usually includes evaluations of, for example, the firm's market position or the quality of its management (Saunders and Allen 2002). The integration of ESG criteria in lending practices would therefore complement traditional credit score systems, providing additional quantitative and qualitative information, thereby improving the reliability of the estimates.

In a credit score system that includes ESG factors, a good borrower is one who pays back his/her debt and also delivers a positive socially responsible performance (Gutiérrez-Nieto et al. 2016). Following this line of reasoning, the integration of ESG screening in lending operations would lead banks to increase the supply and decrease the costs of debt capital for firms engaged in CSR activities. Conversely, for firms with low or no CSR engagement, this approach would imply a higher interest rate or denial of credit. The inclusion of ESG criteria in debt capital allocation processes therefore permits a further distinction between "good" and "bad" borrowers, dictating how the funds are used.

Several studies have in fact empirically demonstrated that a stronger CSR engagement can reduce the cost of debt financing. For example, Sharfman and Fernando (2008), analysing a sample of 267 US firms, find that better management of environmental risks is associated with less expensive debt. The authors argue that good social and environmental performance sends a signal to the financial markets that the firm is less risky, and deserving of lower interest rates. A few years later, Goss and Roberts (2011), investigating whether banks discriminate between borrowers with different levels of CSR, found that firms that presented concerns about their CSR tended to pay higher interest rates, between seven and 18 basis points. Nandy and Lodh (2012) reveal that eco-friendly firms tend to obtain better lending terms than those less engaged in environmental responsibility. Similarly, Jung et al. (2016) document a higher cost of debt for firms more exposed to carbon risk.

Apart from a few exceptions (Hoepner et al. 2016), in which no association is found between borrowers' engagement in CSR and the cost of debt, prior literature largely agrees that CSR has a bearing on the interest rates charged by banks (Banerjee et al. 2019). In a similar way to other extra-financial factors, the exposure to ESG concerns seems to impact on the probability of whether borrowers will repay their loans (i.e. the default risk).

According to Cheng et al. (2014), two main factors explain the superior access to finance conferred on companies more engaged in CSR activities. First, undertaking CSR initiatives favours the building of better relationships with stakeholders, which in turn limits the likelihood of opportunistic behaviour, thereby reducing short-term contracting costs. Second, firms with higher CSR engagement are likely to present a better and more complete disclosure of CSR activities. The higher transparency and reliability of information regarding socially responsible performance limits the information asymmetries between the firm and investors/lenders (Ahn and Choi 2009), thereby reducing the barriers that firms may encounter in accessing funding sources (Diamond and Verrecchia 1991). CSR disclosure helps banks to address their selection problems (Akerlof 1970; Agarwal et al. 2012), permitting them to discriminate more acutely between high- and low-risk borrowers and to apply interest rates consistent with the borrower's profile.

The higher cost of capital for firms with low CSR engagement is likely to impact on the firm's financial performance overall, especially if firms have to constrain or abandon potentially profitable capital investments due to the lack of funds (Scott and Pascoe 1984). To avoid this situation, firms would be well advised to improve their CSR performance.

Just as the presence of SRFs among shareholders may exert an influence on firm orientation in the direction of socially responsible initiatives (Faller and zu Knyphausen-Aufseß 2018), it might be expected that the integration of ESG criteria in banks' lending decisions may stimulate companies to engage in CSR initiatives in order to improve their creditworthiness and reduce the costs of debt financing. As recognised by Scholtens (2009), through credit and equity ownership channels, finance has the potential to promote societal changes by inducing firms to operate in a way that takes account of the ESG implications of their activities.

3.5 ESG and Potential Impacts on Banks' Performance

In parallel with the diffusion of socially responsible initiatives in the banking industry, a fruitful line of research has investigated if and how CSR engagement affects a bank's performance (e.g. Wu et al. 2017). As preliminarily discussed in Sect. 3.2, several factors may explain how the undertaking of socially responsible initiatives can affect the bank's outcomes. A number of recent studies have argued that engagement in socially responsible initiatives may enhance a bank's reputation, with implications for its performance (Wu and Shen 2013; Forcadell and Aracil 2017; Dell'Atti et al. 2017). Drawing on the resource-based view (RBW) perspective (Barney 1991), reputation represents an intangible asset which favours the achievement of competitive advantage (Dowling 2004; Melo and Garrido-Morgado 2012). Academics and researchers have consistently demonstrated that socially responsible activities can improve corporate image and create trusting and long-term relationships with stakeholders, which in turn can lead to better bank performance (McDonald and Lai 2011; Wu and Shen 2013; Wu et al. 2017).

Of all the stakeholder groups, it is customers that appear particularly responsive to CSR programmes (Bhattacharya and Sen 2004; Bhattacharya et al. 2009), with prior studies indicating that CSR has positive effects in terms of customers' responses. For example, Aramburu and Pescador (2019) show that CSR initiatives can positively affect customer loyalty, especially on account of bank reputation as perceived by the customer. Hur et al. (2020), focusing on a sample of 567 South Korean banks, argue that the CSR perception favours the customers' identification, which, in turn, solicits customers' participation to bank CSR initiatives.

Some studies recognise that the undertaking of CSR initiatives may enhance the attractiveness of the firm to job seekers (e.g. Albinger and Freeman 2000; Brammer et al. 2015). Greening and Turban (2000) suggest that because job applicants will likely not have complete information regarding the working environment of their prospective employer, they will rely on the limited information they receive as indicators of the working conditions. The authors claim that CSR engagement can have a positive influence on job applicants' perception of the organisation, thereby increasing its attractiveness as an employer. Other studies demonstrate that CSR activities strengthen an employee's identification with the organisation, reinforcing their commitment (Brammer et al. 2007) and job satisfaction (Valentine and Fleischman 2008).

However, most empirical studies on bank social responsibility have focused on the impact that a bank's engagement in CSR has on financial performance. One of the first studies in this area was by Simpson and Kohers (2002), who noted that a bank's undertaking of socially responsible initiatives has a positive impact on its financial performance. A few years later, Soana (2011), considering a sample of 21 international banks rated by Ethibel, found a positive link between the governance pillar and return on capital invested.

In 2013, Wu and Shen analysed the relationship between bank social responsibility and financial performance, also looking at possible motivations that may induce banks to engage in CSR. The authors find that the relationship between engagement in socially responsible activities and financial performance may be positive, non-negative or non-existent depending on whether CSR programmes are driven by strategic, altruistic or greenwashing motives, respectively. A few years later, using a sample of banks from 18 countries, Shen et al. (2016) showed that socially responsible banks outperform non-socially responsible ones, both in terms of profitability and quality of credit. In the same year, Cornett et al. (2016) examined the relationship between CSR and financial performance of a sample of US commercial banks, over the years of the financial crisis. Their results indicate that larger banks are significantly more engaged in CSR activities than smaller ones.

Wu et al. (2017), having classified a sample of 194 depository-type banks according to their degree of engagement in CSR, demonstrated that the higher the engagement in CSR, the better the bank's profitability and interest margin. This evidence is consistent with that of Forcadell and Aracil (2017), who, examining the European banks listed on the Dow Jones Sustainability Index, demonstrated that a bank's effort to build its reputation through CSR favours the achievement of better economic performance. A few years later, Gangi et al. (2019b) showed that, through the generation and sharing of new knowledge related to CSR, socially responsible banks show a higher marginality and a less risky credit portfolio than less socially responsible ones.

In summary, despite some exceptions that show a neutral link between social and financial performance (e.g. Soana 2011), most of the empirical studies conducted over the last two decades converge in postulating a positive relationship between CSR engagement and economic outcomes in the banking industry.

However, in recent years, authors have seen the necessity of refining investigations into the relationship between CSR and a bank's financial performance by looking at separate areas of CSR in isolation (Miralles-Quirós et al. 2019). As suggested by Melo and Garrido-Morgado (2012) and Gangi et al. (2019b), each separate dimension of CSR taken individually may have a different impact on financial performance. Among the most recent investigations in this regard is the one by Esteban-Sanchez et al. (2017). On the one hand, these authors find that the adoption of CSR practices related to corporate governance and to labour management lead to better financial performance but, on the other, they find that the product responsibility dimension has no effect on a bank's profitability ratios. They also claim that, during the financial crisis period, the undertaking of community social initiatives was a positive predictor of financial performance. Gangi et al. (2019b) show that environmentally friendly banks are less risky than banks with a lower environmental commitment. Similarly, Laguir et al. (2018), analysing a sample of French banks, find that a bank's environmental and financial performances are mutually reinforcing. Finally, Gonenc and Scholtens (2019), focusing on an international sample for the period 2002–2015, find the Tier 1 capital adequacy ratio to be significantly and positively related to ESG indicators. The authors also demonstrate that the relationship may change depending on the period analysed, the institutional context, and a bank's supervisory system.

Despite the growing empirical research in this area, further study still appears necessary (Miralles-Quirós et al. 2019; Herbohn et al. 2019).

Future studies might focus on lending practices, investigating how the integration of specific dimensions of CSR in banks' credit score systems affects economic outcomes. They might also analyse how the impact of CSR engagement on financial performance varies depending on type of bank. For example, it might be illuminating to see if the effects of a bank's undertaking of socially responsible activities change across systemic and local banks.

3.6 Banks' Social Disclosure

For some years now, banks have had an additional means of communication with their stakeholders, the so-called social disclosure or "banks' social disclosure" (BSD). It is broadly defined as non-financial reporting and appears in a variety of manifestations, such as social reports, social statements, sustainability reports and, more recently, integrated reports. The common denominator among these tools is the detection and representation of how the bank integrates social and environmental issues into its business operations.

Perrini (2005) acknowledges the growth of non-financial reporting alongside the evolution of the concept of CSR, contributing to the growing recognition of the importance of companies' accountability. Morsing and Schultz (2006) note that the more organisations make their ethical and social ambitions known, the more likely they are to attract the attention of their stakeholders. Moreover, as it has evolved, CSR has become more unpredictable, changeable and cross-cutting between different industries (Morsing and Schultz 2006), and this includes the banking sector. According to KPMG, who examined the social reporting rate of the banking and financial industry, it now exceeds 70% among banks (KPMG 2017).

The debates about social disclosure must therefore also include banks and can be developed along certain main lines. First, it is necessary to understand the multidimensional nature of the relationship between social performance and social disclosure. Second, it is necessary to consider the evolution of comparability between reporting systems. Third, useful in this regard is an overview of the ESG pillars and the stakeholder categories within BSD.

With reference to the first aspect, CSR initiatives are expensive and not all organisations are able to benefit from this kind of commitment to their stakeholders (Barnett and Salomon 2012), even though stakeholders' capacity to influence (Barnett 2005, 2007) may depend on the strategic use of CSR, including social communication skills. The latter has a triple function: information, control and reputation building (Gangi et al. 2019a). On the one hand, disclosure enables a reduction in asymmetries with stakeholders (Morsing and Schultz 2006); on the other hand, disclosure can be an additional tool for internal control and monitoring of new performance indicators (Vurro and Perrini 2011). Finally, social disclosure can help improve banks' reputations and increase their levels of social legitimacy (Guthrie and Parker 1989; Owen 2008; Gangi and Trotta 2013). The extent of these benefits depends on the social performance that the bank is able to achieve, both within its own organisation and as an influence on other companies.

There are two alternative interpretations of the relationship between BSD and bank social performance (BSP), plus a third that takes a more holistic view (Gangi and D'Angelo 2016). In the first interpretation, BSP is the driver of BSD (Deegan and Blomquist 2006). In other words, banks with better social performance are more incentivised to increase their level of disclosure, thereby dismantling information barriers for stakeholders. However, as Wu and Shen (2013) point out, here we must beware of greenwashing which can occur when a bank wants to appear better than it actually is in terms of its social and environmental compliance. Consistent with Bénabou and Tirole (2010), organisations can vary in the type of commitment they make to CSR—from altruistic to a strategic use of CSR initiatives. As experience grows with time, and as information asymmetries are reduced, in theory stakeholders will be able to discern a bank's true nature.

In the second interpretation, BSD is the driver of BSP. In other words, since BSD is an additional internal control tool, a wider social disclosure offers the opportunity to improve the efficiency and effectiveness of banks' CSR initiatives, as well as the opportunity to better manage relationships with stakeholders (Vurro and Perrini 2011; Michelon et al. 2013).

In the third interpretation, which we find the most convincing, the first two relationships are not mutually exclusive. Rather, BSD and BSP are linked in an interactive relationship (Gangi and D'Angelo 2016). On the one hand, BSP is an incentive to communicate more and in a better way with stakeholders; on the other hand, BSD makes it possible to interpret and improve social performance, contributing to the quality of a bank's commitment to CSR. Once environmental and social KPIs (key performance indicators) have become part of the bank's control dashboard, the reporting system will help to improve internal monitoring and social performance. Within this more holistic perspective, it is important to consider the standardisation of the information in the reporting systems. Over the years, the requirements and standards for acceptable social disclosure have become more stringent. This has helped to bridge the gap in comparability between organisations and their social and environmental certification. UNEP FI makes transparency and accountability its sixth principle. The Global Reporting Initiative (GRI) framework and the dissemination of the UNGC–GRI platform have galvanised the commitment to social communication. Despite some limitations (Pistoni et al. 2018), the Integrating Reporting Framework (<IR>) developed by the International Integrating Reporting Council (IIRC) has given further impetus to the integration of social and environmental data with financial data. International Organization for Standardization (ISO) and Social Accountability (SA) standards have explored specific issues in depth, such as labour (SA 8000), environment (ISO 14001) and social responsibility (ISO 26000). Finally, at the institutional level, several countries are moving to make a commitment to social disclosure more mandatory. For example, the European Union directive 2014/95/EU makes non-financial disclosure mandatory among organisations with over 500 employees. In Italy, Law 254/2016 makes the Consolidated Non-financial Declaration mandatory among large companies. These initiatives are important to the extent that they encourage the adoption of a common and standardised CSR language (Donaldson 1992).

Finally, this third perspective on BSD takes in the idea of broadening social communication, considering the variety of stakeholders to address and interventions that banks can make with respect to the different ESG pillars. The diffusion of ESG criteria among banks has been increasing over time, as demonstrated by the growth in adherence to the Equator Principles and the six principles established by the UNEP FI programme. However, to make the most of BSD, it is important to understand to which pillars banks are most oriented. In order to paint a more complete picture of the role of ESG information in the dialogue between banks and society, in this section we outline the results of a study based on a content analysis of the reporting systems of a panel of the largest commercial banks (by capitalisation) in the EU. The aim is to provide an overview of the orientation of social disclosure on two levels, namely: (i) the relative coverage of the three ESG pillars (see Table 3.3)[4]; and (ii) the distribution of social information with respect to the following main stakeholder categories previously identified in the literature (Perrini 2005; Vurro and Perrini 2011): employees, customers, suppliers and financial partners (see Table 3.4).[5]

It should be noted that across the banks' documents there are variations in reporting systems, according to country (Gangi and Trotta 2013). Social reports are some way from being homogeneous, although the level of standardisation in these documents is increasing. The two largest Italian banks (Intesa and Unicredit) apply Law 254/2016 which prescribes the Consolidated Non-financial Statement based on the GRI standard. The vehicle of Banco Santander, the leading Spanish bank, is its Annual Report. In France, for the two systemic banks BNP Paribas and Credit Agricole it is an Integrated Report, while in Germany Deutsche Bank publishes a Non-financial Report.

3.6.1 BSD Relative to the ESG Pillars

Our analysis (see Table 3.3) shows that the environmental pillar accounts for, on average, 34% of ESG content, with a median value of 32% and a standard deviation of 6%. The social pillar accounts for, on average, 43% of ESG content, with a median value of 45% and a standard deviation of 13%. Finally, the governance pillar accounts for, on average, 23% of ESG content, with a median value of 20% and a standard deviation of 15%. For this pillar, it must be pointed out that the average value is influenced by the weight that governance has in Santander's document, in which, being an annual report, this pillar is reported more thoroughly.

[4] To capture the ESG dimensions, the following keywords were used for English-language documents: *sustainability*, *sustainable*, *environmental*; *social*, *society*; *governance*. For documents in Italian: *sostenibilità*, *ambiente*, *ambientale*; *sociale*, *società*; *governance*. The total (100%) is the sum of all the incidences of words found for each of the three pillars for each country, over the total number of words found for all the three ESG pillars.

[5] The content analysis was carried out using the following keywords: *client/s*, *customer/s* (*clienti* for documents in Italian); *employee/s* (*collaboratori* for documents in Italian); *supplier/s* (*fornitori* for documents in Italian); *shareholder/s*, *investor/s* (*azionisti* or *investitori* for documents in Italian). The total (100%) is the sum of all the incidences of the words found for each stakeholder category, over the total number of words in relation to all categories.

Table 3.3 An analysis of the BSD of leading commercial and systemic banks in the EU, by ESG pillar[a]

Country	*Bank*	*Document*	*Word count by ESG pillars*			
			Environment (%)	*Social (%)*	*Governance (%)*	*Total (%)*
France	BNP Paribas	Integrated Report 2019	32	61	7	100
Spain	Santander	Annual Report 2019	26	20	54	100
Italy	Intesa San Paolo	Consolidated Nonfinancial Statement 2019	30	52	18	100
France	Credit Agricole	Integrated Report 2019	31	50	20	100
Italy	Unicredit	Consolidated Nonfinancial Statement 2019	35	35	29	100
Germany	Deutsche Bank	Non Financial Report 2019	37	39	23	100
Holland	ABN AMBRO	Integrated Annual Review 2019	43	45	11	100
		Average	34	43	23	
		Median	32	45	20	
		Stand. dev.	6	13	15	

[a]Order by capitalization

Source Authors' calculation using data collected from company's online reporting

Table 3.4 An analysis of the BSD of leading commercial and systemic banks in the EU, by stakeholder category[a]

Country	*Bank*	*Document*	*Word count by stakeholder category*				
			Employees (%)	*Clients (%)*	*Suppliers (%)*	*Fin. partners (%)*	*Total (%)*
France	BNP Paribas	Integrated Report 2019	24	59	1	17	100
Spain	Santander	Annual Report 2019	30	32	5	34	100
Italy	Intesa San Paolo	Consolidated Non-financial Statement 2019	23	47	14	15	100
France	Credit Agricole	Integrated Report 2019	22	55	3	20	100
Italy	Unicredit	Consolidated Non-financial Statement 2019	29	53	5	13	100
Germany	Deutsche Bank	Non Financial Report 2019	27	67	0	5	100
Holland	ABN AMBRO	Integrated Annual Review 2019	23	63	4	10	100
		Average	25	54	5	16	
		Median	24	55	4	15	
		Stand. dev.	3	12	5	9	

[a]Order by capitalization

Source Authors' calculation using data collected from company's online reporting

Comparatively, social initiatives occupy more space than the other two pillars. Evidently, then, banks are committed to the communication of their efforts to increase their level of social legitimacy through socially responsible initiatives. Consistent with previous studies (Perrini 2005), we see a strong tendency to describe community relations in terms of local support, corporate giving, donations for art and culture, commitment to social causes in general, including safeguarding of clients and employees.

The second pillar about which banks are communicating is environment (also confirmed by the lower level of standard deviation). This is consistent with the increasing importance that the environment and climate risks have for banks' management, their customer relations and broader risk management. Data and information on the environment are generally provided with regard to the different ways in which sustainable practices can be used to protect natural resources and safeguard future generations. Venturelli et al. (2018, p. 1) point out that the banking industry "has indirect environmental impacts through investing in questionable activities or financing customer who carry out such activities ... A proactive approach to CSR requires the formalization and communication of a strategy that incorporates aspects of sustainability".

Communication on governance issues, however—something that in general affects the financial community to a greater extent—appears to be less substantial. This finding can be traced back to a fragmentation of information regarding this pillar, as reflected in a multitude of banking documents, such as annual financial statements and investor relations reporting (Perrini 2005). Communication on governance is mostly restricted to specific concerns about compliance with self-regulatory systems, shareholder participation and composition, the protection of minority shareholders, and the independence of boards of directors. Nevertheless, the governance pillar should not be underestimated when taking a holistic perspective of the responsibility continuum of banks and organisations in general (Jamali and Sidani 2008; Jamali et al. 2015). Previous studies confirm a positive relationship between effective governance mechanisms and propensity for CSR (Jo and Harjoto 2012; Gangi et al. 2020)—and this includes banks (Gangi et al. 2019b). Moreover, banks can routinely claim superior performance in resolving agency relationships and potential conflicts of interest precisely because of their effective governance mechanisms (Soana 2011; Dincer et al. 2014; Esteban-Sanchez et al. 2017).

Overall, this content analysis based on ESG pillars has confirmed the cross-cutting nature of CSR issues and repeated past studies' findings with reference to other sectors (Perrini 2005). A specific conclusion is that, among banks, the concept of responsible conduct is mainly related to environment and communities. Banks across the board seem to converge on these two issues, which are seen as strategic in their relationships with stakeholders.

3.6.2 BSD and Stakeholder Categories

By adopting a categorisation of BSD consistent with stakeholder theory (Freeman 1984), we can obtain a more complete picture of the recipients of banks' socially responsible initiatives and their CSR-related information (see Table 3.4). An analysis of BSD with respect to banks' stakeholder engagement is important because, as highlighted in prior literature, "banks must disclose information from effective involvement with stakeholders instead of the traditional legitimization and unidirectional communication with stakeholders" (Venturelli et al. 2018, pp. 1–2).

As such, in analysing the social reporting systems of the main systemic and commercial banks in the EU, we see that customers represent the stakeholder category to which, on average, banks are devoting the most information space (54%, with a median value of 55% and a standard deviation of 12%). This evidence is consistent with earlier literature which highlights the opportunity for banks to strategically benefit clients with strong CSR interests (Perez and del Bosque 2013). In fact, through social disclosure, banks have the opportunity to communicate to customers how they can connect with the company by means of products and services that reinforce the social responsibility of the bank, its transparency and its image of environmental responsibility. BSD is also an opportunity to communicate to customers its commitment with regard to the quality of its services and its safeguarding of the interests of the consumer and the community in general.

The strong weighting of "clients" within BSD is consistent with banks' objective of activating an additional communication channel, further to their traditional promotional channels, demonstrating an ability to listen to customers and give prompt responses, under the principle of stakeholder cooperation.

"Employees" are the second-highest category in terms of dedicated space within BSD. The average weight of this stakeholder category is 25%, with a median value of 24% and a very low standard deviation (3%). Socially responsible banks must demonstrate their commitment to employees because they are the ones most directly affected by—or most able to influence—business operations (Perrini 2005; Dangelico 2015). Also, banks have an interest in promoting an image of excellence in meeting employee expectations in order to increase their attractiveness as an employer and retain the best human resources (Barrena-Martinez et al. 2018). Initiatives reported on may include: concern for the workplace, workers' health and safety, effective internal communication, workers' rights, transparency, and equal opportunities—all initiatives that can serve to improve the working environment and workers' quality of life (Perrini 2005). There are beneficial effects of this in financial terms, too, as evidenced by several studies on the positive correlation between the banks' quality of relationships with their employees and their financial performance (Mention and Bontis 2013; Esteban-Sanchez et al. 2017).

"Financial partners" represent the next-largest category, with the average among the main EU systemic banks being 16%, with a median value of 15%. Financial partners are important in establishing a bank's identity, values and mission. However, as highlighted in previous studies (Perrini 2005)—and consistent with the above evidence on governance information as an ESG pillar—communication to shareholders and other investors is not given much space in social reporting; governance and investor relations issues are fragmented across a variety of documents such as financial statements, financial reports and investor relations reports. Nonetheless, financial partners are stakeholders, too, and this fact becomes more important from a CSR perspective, especially if we consider those that have a commitment to the bank's social responsibility and the mainstreaming of socially responsible investing. BSD aimed at financial partners is an area worthy of further development since it could help demonstrate a bank's ability to combine the creation of social value with the creation of shareholder value.

Finally, we come to "suppliers", who occupy an average of 5% of communication space, with a median of 4% and a low standard deviation (5%). This low figure probably reflects the operational characteristics of the banking industry, for which socially responsible functioning is inherently less tied to supply chains than other industries, such as manufacturing. However, this category could also usefully expand in future BSD, to the extent that a bank is socially responsible if it is able to establish better relationships with all the various stakeholders with which it interacts and which support its operations. The quality of a bank's operating model must also depend to some degree on the social and environmental responsibility of its suppliers, including those who design socially responsible financial products and services.

Overall, these analyses have helped us to identify the ESG pillars to which socially responsible banks are most committed (environment and society) and the stakeholder categories with which banks are engaging with the most in their information exchange activities (customers and employees). For one thing, these findings confirm that systemic banks are adopting a strategic and selective approach to CSR (Wu and Shen 2013); secondly, there are potential areas for expansion within these social disclosure channels (e.g. on governance and investor relations).

3.7 Ethical Versus Conventional Banks: An Explorative Comparative Analysis

There has been a significant evolution of CSR in the financial system in recent years, with financial institutions increasingly being required to integrate ESG criteria into their credit screening process to adequately respond to the needs of stakeholders. Banks have traditionally played a central role in the economy, on account of their exclusive role as analysts of and consultants for the exchange of financial resources and the creation of value between businesses and companies. With the 2008 crisis, however, discourse on the sustainability of banks and ESG integration began to follow an accelerated and more urgent course, because, as noted by the literature and the regulators, these institutions failed in their social responsibilities by focusing solely on maximising shareholder value and their own profits, utterly ignoring the consequences of their investment choices on financial and economic systems. In failing to fulfil their responsibilities to society, banks missed their opportunity to play a proactive role in creating financial products and markets that generate value for all market players, from business to society. So the recent emergence of ethical questions being asked of the financial system has its roots in society's expectations of greater responsibility from these actors in the generation of collective wealth. The loss of trust felt by consumers, and by stakeholders in general, created sufficient space in the financial market for the emergence of so-called sustainable banks (Scholtens 2009; Cowton 2002; Paulet et al. 2015; Valls Martínez et al. 2020). In the aftermath of the financial crises, stakeholders became more inclined to see a role for social and environmental criteria alongside financial ones, whereupon certain banking operators capitalised on this movement by developing financial products and credit policies that catered for profit maximisation expectations for both business and society. As Valls Martínez et al. (2020) point out, the sustainable banking concept encompasses different operators, among them savings and cooperative banks, as well as ethical banks. Following Cowton (2002), the three main characteristics that define an ethical bank are integrity, responsibility and affinity. Integrity refers to the ability to generate the trust among savers and borrowers that is needed to ensure sustainable development; responsibility means taking on board the implications and consequences of lending policies; and affinity defines the ability of the ethical banking business model to attune to the financial needs of its borrowers, understanding customers' and depositors' preferences in terms of their investments. Ethical banks are therefore distinguished by a more rigid approach with regard to their integration of social and environmental criteria into the whole investment policy and credit business model. Despite the fact that conventional and ethical banks both operate in the credit segment, collect money from the markets and lend it to organisations, families and individuals, the two types differ significantly in terms of their business models (Cowton 2002; Paulet et al. 2015).

The mission of ethical banks sees them addressing the original function of the financial market: that is, creating the right conditions for capital seekers and investors in which to cooperate in the furtherance of sustainable development. As such, their credit activity is heavily oriented towards the creation of a positive impact from all three perspectives: social, environmental and economical (Paulet et al. 2015). Their ethical approach has a dual nature. First, they generally refrain from engaging with controversial businesses, such as the defence industry, highly polluting sectors, or grey business related to illegal activities. Second, they see it as their purpose to raise public awareness of the role of capital in supporting a resilient economy that can invest in long-term horizons. Through their activities, ethical banks promote the idea of financial inclusion, which serves to strengthen solidarity and support social justice in pursuit of sustainable development based on environmental protection and fairer societies. These banks are generally associated with a higher percentage of ESG-oriented organisations that meet social and environmental criteria, and this implies that they see social and environmental values as being inseparable from economic impact. In fact, ethical banks' mission statements are likely to make explicit their integration of financial considerations with social and environmental assessment (Paulet et al. 2015; Valls Martínez et al. 2020). This enhanced ability to integrate multiple dimensions within economic sustainability is generally associated with a lower level of operational risk. Specific reasons can be identified to explain ethical banks' potential enhanced long-term resilience compared to conventional banks.

First, they differ from conventional or systemic banks in that they base their lending business on savers' and customers' capital, reinvesting it in activities and businesses promoting social and environmental values. From this perspective, the role of an ethical bank is one of supporting the common good by linking customer capital to businesses oriented towards sustainable development, thus creating a virtuous circle of capital generated and invested in real economies (Paulet et al. 2015). Furthermore, ethical banks are less willing than traditional ones to look at the financial market or to engage in speculative investment practices. This category of bank, then, has its presence largely in the real economy, rather than in the globalised financial market. In terms of economic sustainability, ethical banks' profits are generally reinvested mostly in the banks' own social missions, and they are also more willing than conventional banks to support projects with a high financial risk (Paulet et al. 2015).

The second distinguishing feature of an ethical bank is its stakeholder governance model. Ethical banks are generally heavily involved in local social and entrepreneurial communities, where they are active players in those economic contexts. Unlike conventional banks, they tend to engage with their customers in the form of partnerships, integrating them, and their needs, into their business models and actions (Paulet et al. 2015). The stronger relationships between ethical banks and their stakeholders are generally attributed to these banks' higher degree of transparency about their activities. Next to communication, transparency is a core value of an ethical bank's business model (Cowton 2002). These higher levels of transparency and participation by stakeholders in the bank's decisions, along the whole financial chain, reinforce, in a virtuous circle, stakeholders' awareness and support, as well the banks' capacity to act in accordance with their tripartite business model (economic–social–environmental sustainability). Further, conventional banks are, in general, highly globalised, with a greater number of operative branches than ethical banks (Valls Martínez et al. 2020).

Ethical banks and ethical finance are often presented as an alternative model which, alongside financial considerations, embraces values such as solidarity, trust, cooperation and transparency (Paulet et al. 2015; Valls Martínez et al. 2020); however, the literature as yet offers few insights into how these kinds of bank perform relative to conventional ones (Paulet et al. 2015; Banca Etica 2017; Valls Martínez et al. 2020). Paulet et al. (2015), analysing the features that distinguish ethical banks from conventional ones, conclude that ethical banks have remained largely unchanged as a result of the new post-financial-crisis landscape and the financial regulations that accompanied it. Valls Martínez et al. (2020) analyse the economic structure of ethical banks by comparing a European ethical bank with two conventional ones, and empirically demonstrate that ethical banks, although not outperforming conventional ones, tend to be more stable.

This section offers an explorative comparative analysis of ethical and conventional banks' performance, in terms of profitability and quality of credit policy. Ethical banks represent a small percentage of the banking sector as a whole, which goes towards explaining the scarcity of data available. In fact, no hallmarks or certifications have been devised to formally recognise ethical banks and distinguish them from conventional ones. However, there is a global movement aimed at promoting positive ethical change in the industry. The Global Alliance for Banking on Values (GABV) is a global banking network committed to promoting a viable banking system, founded on a framework of five principles: (1) socially and environmentally sustainable projects which are (2) grounded in local communities; (3) the promotion of long-term relationships with clients; (4) to be self-sustaining and resilient to outside disruptions; and (5) transparent and inclusive governance mechanisms. As of 2020, the GAVB consisted of 63 banking entities operating in regions across the world (Asia, Africa, Australia, Latin America, North America and Europe).

The ethical banks among the GAVB network membership constituted the initial line of enquiry in our explorative analysis. GAVB members operate mostly in Europe (26%) and the US and Canada (24%). Given the relevance of SRI investments in both Europe and the US (GSIA 2018), we chose to limit our screening to those areas. Subsequently, we consulted further sources for screening for ethical banks (Paulet et al. 2015; Banca Etica 2017). By combining Banca Etica's report on sustainable banking (2017) and Paulet et al.'s study (2015) with the GAVB member list, we detected eight ethical and eight conventional banks, based both in Europe and the US, making a total of 16 banking operators selected. Table 3.5 lists the sample of banks by country.

The financial performance of the 16 banks was gathered from Thomson Reuters Datastream where possible (81% of cases); otherwise, data were manually retrieved from financial statements published by banks (19% of observations). We collected banks' financial information for an 11-year time horizon, from 2009 to 2019.

Table 3.5 Institutions included in the study

Conventional	*Ethical*	*Country*
Banca Intesa	Banca Etica	Italy
Santander	CaixaBank	Spain
Jyske Bank	Merkur Cooperative Bank	Denmark
Commerzbank	UnweltBank	Germany
ING Bank	Triodos Bank	Netherlands
Novascozia	Kindred Credit Union	Canada
Eastern Bank	PNC Bank	USA
CityBank	U.S. Bancorp	USA

Source Data collected from Banca Etica report (2017), Paulet et al. (2015) and GAVB network

In order to identify potential differences between ethical and conventional banks, in line with prior studies (Paulet et al. 2015; Valls Martínez et al. 2020), the explorative analysis used a set of parameters that captured the stability, solvency and profitability of banks' business models. More specifically, bearing in mind ethical banks' preference for being involved with real economies rather than financial markets, a first proxy employed was the relationship between total activities and the level of a bank's loans. In line with prior studies (Valls Martínez et al. 2020; Banca Etica 2017), the higher the ratio, the lower a bank's involvement in financial markets and the more oriented it is towards being active in the real economy (a greater proportion of loans to assets). As such, we would expect ethical banks to be the ones showing a higher proportion of loans to total assets. Secondly, the literature recognises that ethical banks mostly rely on customers' and savers' deposits for capital, while conventional banks make greater use of other sources of capital, such as the financial markets. Thus, a second measure adopted is the ratio between total deposits and total bank liabilities as a proxy for the bank's capital acquisition strategy. The higher this ratio, the more a bank relies on capital from the real economy to invest or to lend; the lower this ratio, the higher the proportion of a bank's capital sources from financial markets. As regards solvency, the measure adopted as a proxy for banks' credit quality is loan loss provision to total loans (LLP). The LLP ratio measures loan quality, this being the outcome of banks' financial activities, which reflects the ratio of arrears, insolvent loans and loan losses after the reclamation of possible collateral. In order to evaluate banks' stability, the study considered the ratio between net equity and total liabilities, and the level of Tier 1 as a key

Table 3.6 Description of variables

Variable	*Description*
Loans_totAsset	Total loans to total asset
Deposit_totLiab	Total deposits over total liabilities
Equity_totLiab	Net equity over total liabilities
Tier_1%	Tier 1 capital (%)
LLP	Loan loss provision to total loans
NIM	Net interest margin
ROA	Return on assets
ROE	Return on equity

measure of a bank's financial strength and its level of capitalisation. Finally, in line with prior studies (Valls Martínez et al. 2020), the profitability indicator considered was the net interest margin (NIM), representing the difference between interest income generated and total amount of interest paid to lender, which was followed by an approximation of the likelihood of a bank to thrive in the long run (Table 3.6).

Table 3.7 presents descriptive statistics for the two groups of banks. As can be seen, the volume of loans to total assets is higher for the ethical banks over the period analysed. Contrary to what was expected, the ratio

Table 3.7 Credit quality of ethical and conventional banks: univariate analysis

Variables (2009–2019)	*Obs.*	*Ethical*	*Conv.*	*Ethical*	*Conv.*	*Ethical*	*Conv.*	*MW test*	
Total obs.	*176*	*Mean*		*Stand. dev.*		*Median*		*(z_score)*	
Loans_totAsset	169	0.58	0.54	0.26	0.10	0.62	0.55	−4.189	***
Deposit_totLiab	169	0.61	0.65	0.37	0.16	0.75	0.67	−1.718	*
Equity_totLiab	169	0.11	0.09	0.14	0.03	0.07	0.08	0.950	
Tier_1%	137	11.36	12.65	3.49	2.14	10.08	12.50	3.210	***
LLP	71	0.0079	0.0074	0.0060	0.0055	0.0050	0.0053	3.736	
NIM	143	4.94	2.30	11.71	0.84	2.97	2.37	−3.083	***
ROA	127	0.67	0.62	0.59	0.44	0.40	0.60	0.787	
ROE	137	7.18	7.94	4.61	5.58	5.36	7.85	1.424	

Note This table shows the medians of the banks' financial indicators and the results of the Mann–Whitney test. The variables compared are: Loans over total assets, Total deposits over total liabilities, Tier 1, NIM, ROA and ROE. *, ** and *** indicate statistical significance at 10%, 5% and 1% levels, respectively

Source Authors' calculations using data collected from Thomson Reuters Datastream and banks' financial statements

of deposits to total liabilities is actually higher for conventional banks. In terms of stability indicators, the level of net equity over total liabilities is on average higher for ethical banks, while Tier 1 capital is greater in conventional banks. With reference to the profitability ratio, ethical banks in the period analysed demonstrated a higher NIM and ROA, while ROE is generally higher for conventional banks.

In summary, from our descriptive analysis we observed several differences between conventional and ethical banks. In order to test their statistical significance, we applied a non-parametric test, namely a between-sample analysis using the Mann–Whitney test, on account of the relatively small sample size. The Mann–Whitney test indicated significant differences among the median of indicators used to investigate the two groups of banks.

In particular, over the time horizon studied, ethical banks reveal significantly higher median values for the loans to total assets ratio than do conventional ones (at a 1% significance level), analogous results emerge for NIM (at a 1% significance). Moreover, we observe that the median of deposit over total liabilities is higher (0.75) for ethical banks relative to conventional ones (0.67) at 10% of significance, while the Tier 1 level, showed significantly higher median values for conventional banks (at 1% significance level).

Taken as a whole, these findings are consistent with prior literature. First, with reference to ethical banks' business models, our analysis confirms that, in general, the level of loans over total assets is noticeably higher in ethical banks, thus indicating that these banks are deeply engaged in their function as intermediaries between those who need capital and those who are willing to invest in the real economy. As regards source of capital, ethical banks emerge from our analysis as demonstrating a higher level of deposit over total liabilities. This finding might be a consequence of the efforts that have been made by ethical banks by collecting more capital from depositors and less from financial markets (Paulet et al. 2015; Valls Martínez et al. 2020). In terms of stability indicators, results from the non-parametric test show that conventional banks tend to evince a statistically significant higher median value for the Tier 1 capital ratio; however, this might be a result of the more stringent patrimonial requirements requested imposed on systemic banks after the financial crisis. Finally, the business model of ethical banks is rewarded in terms of higher interest margin. This is a good signal, especially during a period of very low interest rate on capital markets.

References

Ahmed SU, Islam ZM, Hasan I (2012) Corporate social responsibility and financial performance linkage evidence from the banking sector of Bangladesh. J Manag Organ 1(1):14–21

Ahn S, Choi W (2009) The role of bank monitoring in corporate governance: evidence from borrowers' earnings management behaviour. J Bank Financ 33(2):425–434. https://doi.org/10.1016/j.jbankfin.2008.08.013

Agarwal S, Chang Y, Yavas A (2012) Adverse selection in mortgage securitization. J Financ Econ 105(3):640–660. https://doi.org/10.1016/j.jfineco.2012.05.004

Akerlof G (1970) The market for lemons: qualitative uncertainty and the market mechanism. Q J Econ 84(3)

Albinger HS, Freeman SJ (2000) Corporate social performance and attractiveness as an employer to different job seeking populations. J Bus Soc Rev 28:243–253. https://doi.org/10.1023/a:1006289817941

Aramburu IA, Pescador IG (2019) The effects of corporate social responsibility on customer loyalty: the mediating effect of reputation in cooperative banks versus commercial banks in the Basque country. J Bus Soc Rev 154:701–719. https://doi.org/10.1007/s10551-017-3438-1

Asrar-ul-Haq M, Kuchinke KP, Iqbal A (2017) The relationship between corporate social responsibility, job satisfaction, and organizational commitment: Case of Pakistani higher education. J Clean Prod 142(4):2352–2363. https://doi.org/10.1016/j.jclepro.2016.11.040

Banca Etica (2017) Annual report 2017

Bank of Italy (2020) Annual report 2020

Banerjee A, Niehaus P, Suri T (2019) Universal basic income in the developing world. Annu Rev Econom 11:959–983. https://doi.org/10.1146/annurev-economics-080218-030229

Barnea A, Rubin A (2010) Corporate social responsibility as a conflict between shareholders. J Bus Soc Rev 97:71–86. https://doi.org/10.1007/s10551-010-0496-z

Barnett ML (2005). Stakeholder influence capacity and the variability of financial returns to corporate social responsibility. Proc Int Assoc Bus Soc 16:287–292. https://doi.org/10.5465/amr.2007.25275520

Barnett ML, Salomon RM (2006) Beyond dichotomy: the curvilinear relationship between social responsibility and financial performance. Strateg Manag J 27 (11):1101–1122. https://doi.org/10.1002/smj.557

Barnett ML (2007) Stakeholder influence capacity and the variability of financial returns to corporate social responsibility. Acad Manage Rev 32(3):794–816. https://doi.org/10.5465/amr.2007.25275520

Barnett ML, Salomon RM (2012) Does it pay to be really good? Addressing the shape of the relationship between social and financial performance. Strateg Manag J 33(11):1304–1320. https://doi.org/10.1002/smj.1980

Barney J (1991) Firm resources and sustained competitive advantage. J Manage 17(1):99–120. https://doi.org/10.1177/014920639101700108

Barrena-Martinez J, López-Fernández M, Romero-Fernandez PM (2018) Drivers and barriers in socially responsible human resource management. Sustainability 10(5):1532. https://doi.org/10.3390/su10051532

Bénabou R, Tirole J (2010) Individual and corporate social responsibility. Economica 77(305):1–19. https://doi.org/10.1111/j.1468-0335.2009.00843.x

Bencivenga VR, Smith BD (1993) Some consequences of credit rationing in an endogenous growth model. J Econ Dyn Control 17(1):97–122. https://doi.org/10.1016/s0165-1889(06)80006-0

Bhattacharya CB, Sen S (2004) Doing better at doing good: when, why, and how consumers respond to corporate social initiatives. Calif Manage Rev 47(1):9–24. https://doi.org/10.2307/41166284

Bhattacharya CB, Korschun D, Sen S (2009) Strengthening stakeholder–company relationships through mutually beneficial corporate social responsibility initiatives. J Bus Soc Rev 85:257–272. https://doi.org/10.1007/s10551-008-9730-3

Brammer S, Millington A (2003) The effect of stakeholder preferences, organizational structure and industry type on corporate community involvement. J Bus Soc Rev 45:213–226. https://doi.org/10.1023/a:1024151528646

Brammer S, Millington A, Rayton B (2007) The contribution of corporate social responsibility to organizational commitment. Int J Hum Resour Manag 18(10):1701–1719. https://doi.org/10.1080/09585190701570866

Brammer S, He H, Mellahi K (2015) Corporate social responsibility, employee organizational identification, and creative effort: the moderating impact of corporate ability. Group Organ Manag 40(3):323–352. https://doi.org/10.1177/1059601114562246

Calderon F, Chong LC (2014) Dilemma of sustainable lending. J Sustain Finance Invest 4(2):192–209. https://doi.org/10.1080/20430795.2014.883302

Cai L, Cui J, Jo H (2016) Corporate environmental responsibility and firm risk. J Bus Soc Rev 139:563–594. https://doi.org/10.1007/s10551-015-2630-4

Calton JM, Payne SL (2003) Coping with paradox: Multistakeholder learning dialogue as a pluralist sensemaking process for addressing messy problems. Bus Soc 42(1):7–42. https://doi.org/10.1177/0007650302250505

Capra F, Stewart J, Reed D, Barrymore L (1946) It's a wonderful life. RKO

Carvalho JMS, Jonker J, Dentchev N (2014) What's in a word? An exploration of the changes in meaning of corporate social responsibility over the last century with an emphasis on the last decades. In D. Turker, H. Toker, & C. Altuntas (eds) Contemporary issues in corporate social responsibility. Lexington Books

Cheng B, Ioannou I, Serafeim G (2014) Corporate social responsibility and access to finance. Strateg Manag J 35(1), 1–23. https://doi.org/10.1002/smj.2131

Chemmanur TJ, Fulghieri P (1994) Investment bank reputation, information production, and financial intermediation. J Finance 49(1):57–79. https://doi.org/10.1111/j.1540-6261.1994.tb04420.x
Consob (2020) Bollettino Statistico—2020
Cornett MM, Erhemjamts O, Tehranian H (2016) Greed or good deeds: An examination of the relation between corporate social responsibility and the financial performance of US commercial banks around the financial crisis. J Bank Financ 70:137–159. https://doi.org/10.1016/j.jbankfin.2016.04.024
Costello AM, Wittenberg-Moerman R (2011) The impact of financial reporting quality on debt contracting: evidence from internal control weakness reports 49(1):97–136. https://doi.org/10.1111/j.1475-679x.2010.00388.x
Cowton CJ (2002) Integrity, responsibility and affinity: three aspects of ethics in banking. Bus Soc Rev 11(4):393–400. https://doi.org/10.1111/1467-8608.00299
Cox P, Brammer S, Millington A (2004) An empirical examination of institutional investor preferences for corporate social performance. J Bus Soc Rev 52:27–43. https://doi.org/10.1023/b:busi.0000033105.77051.9d
Cox P, Brammer S, Millington A (2008) Pension funds and corporate social performance: An empirical analysis. Bus Soc 47(2):213–241. https://doi.org/10.1177/0007650306297945
Curado C (2008) Perceptions of knowledge management and intellectual capital in the banking industry. J Knowl Manag 12(3):141–155. https://doi.org/10.1108/13673270810875921
Dalton DR, Hitt MA, Certo T, Dalton CM (2007) The fundamental agency problem and its mitigation. Acad Manage Rev 1(1):1–64. https://doi.org/10.5465/078559806
Dam L, Scholtens B (2012) Does ownership type matter for corporate social responsibility? Corp Gov 20(3):233–252. https://doi.org/10.1111/j.1467-8683.2011.00907.x
Dangelico RM (2015) Improving firm environmental performance and reputation: The role of employee green teams. Bus Strategy Environ 24(8):735–749. https://doi.org/10.1002/bse.1842
De Andres P, Vallelado E (2008) Corporate governance in banking: The role of the board of directors. J Bank Financ 32(12):2570–2580. https://doi.org/10.1016/j.jbankfin.2008.05.008
de Bruin B (2015) Ethics and the global financial crisis: why incompetence is worse than greed. Cambridge: Cambridge University Press
Decker OS (2004) Corporate social responsibility and structural change in financial services. Manag Audit J 19(6):712–728. https://doi.org/10.1108/02686900410543840
Deegan C, Blomquist C (2006) Stakeholder influence on corporate reporting: an exploration of the interaction between WWF-Australia and the Australian

minerals industry. Account Organ Soc 31(4):343–372. https://doi.org/10.1016/j.aos.2005.04.001
Dell'Atti S, Trotta A, Iannuzzi AP, Demaria F (2017) Corporate social responsibility engagement as a determinant of bank reputation: an empirical analysis. Corp Soc Responsib Environ Manag 24(6):589–605. https://doi.org/10.1002/csr.1430
Déniz-Déniz MDLC, Saá-Pérez D (2003) A resource-based view of corporate responsiveness toward employees. Organ Stud 24(2):299–319. https://doi.org/10.1177/0170840603024002347
Dhaliwal DS, Li OZ, Tsang A, Yang YG (2011) Voluntary nonfinancial disclosure and the cost of equity capital: the initiation of corporate social responsibility reporting 86(1):59–100. https://doi.org/10.2308/accr.00000005
Diamond DW, Verrecchia RE (1991) Disclosure, liquidity, and the cost of capital. J Finance 46(4):1325–1359. https://doi.org/10.1111/j.1540-6261.1991.tb04620.x
Dimson E, Karakaş O, Li X (2015) Active ownership. Rev Financ Stud 28(12):3225–3268. https://doi.org/10.1093/rfs/hhv044
Dincer H, Celik IE, Yilmaz R, Hacioglu Ü (2014) The financial implications of corporate social responsibility in the banking sector. In Managerial issues in finance and banking (pp. 197–207). Springer, Cham
Dombret AR, Kern HJ (2003) European retail banks—an endangered species? Monitor Group, Rothschild, Wiley, Frankfurt am Main, Chichester, New York (NY)
Donaldson T (1992) The language of international corporate ethics. In Business ethics: Japan and the global economy (pp. 115–131). Springer, Dordrecht
Dowling GR (2004) Corporate reputations: should you compete on yours? Calif Manage Rev 43(3):19–36. https://doi.org/10.2307/41166219
Drucker PF (1993) The rise of the knowledge society. Wilson Quarterly 17(2):52–71
Du X, Weng J, Zeng Q, Chang Y, Pei H (2017) Do lenders applaud corporate environmental performance? Evidence from Chinese private-owned firms. J Bus Soc Rev 143(1):179–207. https://doi.org/10.1007/s10551-015-2758-2
Dyck A, Lins KV, Roth L, Wagner HF (2019) Do institutional investors drive corporate social responsibility? International evidence. J financ econ 131(3):693–714. https://doi.org/10.1016/j.jfineco.2018.08.013
Einaudi L (1921) Gli ideali di un economista. Soc. an. ed. "La Voce"
El Ghoul S, Guedhami O, Kwok CCC, Mishra DR (2011) Does corporate social responsibility affect the cost of capital? J Bank Financ 35(9):2388–2406. https://doi.org/10.1016/j.jbankfin.2011.02.007
Elkington J (1997) The triple bottom line. Environmental management: Readings and cases

Ernst & Young (2019) EY global review—2019

Esteban-Sanchez P, de la Cuesta Gonzalez M, Paredes Gazquez JD (2017) Corporate social performance and its relation with corporate financial performance: International evidence in the banking industry. J Clean Prod 162:1102–1110. https://doi.org/10.1016/j.jclepro.2017.06.127

Esty DC, Levy MA, Srebotnjak T, de Sherbinin A (2005) Environmental sustainability index: benchmarking national environmental stewardship. Yale Center for Environmental Law and Policy, New Haven

European Banking Authority (2019) EBA Action Plan on Sustainable Finance 2019

European Banking Federation (2008) Annual report

Eurosif (2018) European SRI study

Faller CM, zu Knyphausen-Aufseß D (2018) Does equity ownership matter for corporate social responsibility? A literature review of theories and recent empirical findings. J Bus Soc Rev 150:15–40. https://doi.org/10.1007/s10551-016-3122-x

Financial Crisis Inquiry Commission (2011) Financial Crisis Inquiry Commission Report—2011

Forcadell FJ, Aracil E (2017) european banks' reputation for corporate social responsibility 24(1):1–14. https://doi.org/10.1002/csr.1402

Freeman RE (1984) Strategic management: a stakeholder theory. J Manage Stud 39(1):1–21

Friedman M (1970) The social responsibility of business is to increase its profits. The New York Times Magazine, New York, NY, 13 September

Gangi F, Trotta C (2013) Determinants of corporate social disclosure through a multi-perspective approach: evidences from the target companies of socially responsible funds. Accounting and Control for Sustainability (Studies in Managerial and Financial Accounting, Vol. 26), Emerald Group Publishing Limited, 33–75. https://doi.org/10.1108/s1479-3512(2013)0000026002

Gangi F, D'Angelo E (2016) The virtuous circle of corporate social performance and corporate social disclosure. Mod Econ 7:1396–1418. https://doi.org/10.4236/me.2016.712129

Gangi F, Mustilli M, Varrone N, Daniele LM (2018) Corporate social responsibility and banks' financial performance. Int Bus Res 11(10). https://doi.org/10.5539/ibr.v11n10p42

Gangi F, Méric J, Jardat R, Daniele LM (2019a) Business for society. Routledge

Gangi F, Mustilli M, Varrone N (2019b) The impact of corporate social responsibility (CSR) knowledge on corporate financial performance: evidence from the European Banking industry. J Knowl Manag 23(1):110–134

Gangi F, Daniele LM, Varrone N (2020) How do corporate environmental policy and corporate reputation affect risk-adjusted financial performance? Bus Strategy Environ 29(5):1975–1991. https://doi.org/10.1002/bse.2482

Gonenc H, Scholtens B (2019) Responsibility and performance relationship in the banking industry. Sustainability 11(12):3329. https://doi.org/10.3390/su11123329

Goss A, Roberts GS (2011) The impact of corporate social responsibility on the cost of bank loans. J Bank Financ 35(7):1794–1810. https://doi.org/10.1016/j.jbankfin.2010.12.002

Goyal KA, Joshi V (2011) A study of social and ethical issues in banking industry. Journal of Economics & Research 2(5):49–57

Graves SB, Waddock SA (1994) Institutional owners and corporate social performance. Acad Manage J 37(4). https://doi.org/10.5465/256611

Greenbaum SI, Thakor AV (2007) Contemporary financial intermediation, 2nd edn. Academic Press, Amsterdam

Greening DW, Turban DB (2000) Corporate social performance as a competitive advantage in attracting a quality workforce. Bus Soc 39(3):254–280. https://doi.org/10.1177/000765030003900302

Greenwood J, Jovanovic B (1990) Financial development, growth, and the distribution of income. J Polit Econ 98(5):1076–1107

Global Sustainable Investment Alliance (GSIA) (2018) Global Sustainable Investment Review 2018

Gupta K (2018) Environmental sustainability and implied cost of equity: International evidence. J Bus Soc Rev 147:343–365. https://doi.org/10.1007/s10551-015-2971-z

Gutiérrez-Nieto B, Serrano-Cinca C, de la Cuesta González M (2016) A multivariate study of over-indebtedness' causes and consequences. Int J Consum Stud 41(2):188–198. https://doi.org/10.1111/ijcs.12324

Guthrie J, Parker LD (1989) Corporate social reporting: a rebuttal of legitimacy theory. Account Bus Res 19(76):343–352. https://doi.org/10.1080/00014788.1989.9728863

Haigh M, Hazelton J (2004) Financial markets: a tool for social responsibility? J Bus Soc Rev 52:59–71. https://doi.org/10.1023/b:busi.0000033107.22587.0b

Hancock J (2005) Investing in corporate social responsibility: a guide to best practice. Business Planning & the UK's Leading Companies, 25–38

Harjoto MA, Jo H (2011) Corporate governance and CSR nexus. J Bus Soc Rev 100:45–67. https://doi.org/10.1007/s10551-011-0772 6

Herbohn K, Gao R, Clarkson P (2019) Evidence on whether banks consider carbon risk in their lending decisions. J Bus Soc Rev 158:155–175. https://doi.org/10.1007/s10551-017-3711-3

Hoepner A, Oikonomou I, Scholtens B, Schröder M (2016) The effects of corporate and country sustainability characteristics on the cost of debt: an international investigation. J Bus Finance Account 43(1):158–190. https://doi.org/10.1111/jbfa.12183

Hong H, Sraer D (2011) Quiet bubbles. Working Paper Princeton University, Princeton

Hook L (2018) Central bank chiefs sound warning on climate change. The Financial Times, April 2018

Hur WM, Moon TW, Kim H (2020) When and how does customer engagement in CSR initiatives lead to greater CSR participation? The role of CSR credibility and customer–company identification. Corp Soc Responsib Environ Manag 27(4):1878–1891. https://doi.org/10.1002/csr.1933

Il Sole 24 Ore (2011) Il vero conto del salvataggio delle banche americane. Accessed 2 December 2011

International Monetary Fund (2018) International Monetary Fund Annual Report—2018

Jamali D, Sidani Y (2008). Classical vs. modern managerial csr perspectives: insights from Lebanese context and cross cultural implications. Bus Soc Rev 113 (3):329–346. https://doi.org/10.1111/j.1467-8594.2008.00323.x

Jamali D, Safieddine AM, Rabbath M (2008) Corporate governance and corporate social responsibility synergies and interrelationships. Corp Gov 16(5):443–459. https://doi.org/10.1111/j.1467-8683.2008.00702.x

Jamali D, El Dirani AM, Harwood IA (2015) Exploring human resource management roles in corporate social responsibility: the CSR-HRM co-creation model. Bus Ethics 24(2):125–143. https://doi.org/10.1111/beer.12085

Jensen MC, Meckling WH (1976) Theory of the firm: managerial behavior, agency costs and ownership structure. J financ econ 3(4):305–360. https://doi.org/10.1016/0304-405x(76)90026-x

Jeucken MHA, Bouma JJ (1999) The changing environment of banks. Greener Manag Int 99(27):21

Jia M, Zhang Z (2013) The CEO's representation of demands and the corporation's response to external pressures: do politically affiliated firms donate more? Manag Organ Rev 9(1):87–114. https://doi.org/10.1111/j.1740-8784.2012.00297.x

Jo H, Harjoto MA (2011) Corporate governance and firm value: the impact of corporate social responsibility. J Bus Soc Rev 103:351–383. https://doi.org/10.1007/s10551-011-0869-y

Jo H, Harjoto MA (2012) The causal effect of corporate governance on corporate social responsibility. J Bus Ethics 106(1):53–72

Jo H, Kim H, Kwangwoo P (2015) Corporate environmental responsibility and firm performance in the financial services sector. J Bus Ethics, 131:257–284. https://doi.org/10.1007/s10551-014-2276-7

Johnsen DB (2003) Socially responsible investing: a critical appraisal. J Bus Soc Rev 43:219–222. https://doi.org/10.1023/a:1022998232503

Jung J, Herbohn K, Clarkson P (2016) Carbon risk, carbon risk awareness and the cost of debt financing. J Bus Ethics 150(4):1151–1171. https://doi.org/10.1007/s10551-016-3207-6
King RG, Levine R (1993) Finance and growth: Schumpeter might be right. Q J Econ 108(3):717–737. https://doi.org/10.2307/2118406
Kolk A (2016) The social responsibility of international business: from ethics and the environment to CSR and sustainable development. J World Bus 51(1):23–34. https://doi.org/10.1016/j.jwb.2015.08.010
KPMG Corporate Responsibility Reporting in the Banking Sector 2017
Laguir I, Marais M, El Baz J, Stekelorum R (2018) Reversing the business rationale for environmental commitment in banking. Manag Decis 56(2):358–375. https://doi.org/10.1108/md-12-2016-0890
Lentner C, Szegedi K, Tatay T (2015) Corporate social responsibility in the banking sector. Public Finance Quarterly (1):95–103
Levine R (2005) Finance and growth: theory and evidence. Handb Econ Growth 1:865–934. https://doi.org/10.1016/s1574-0684(05)01012-9
Levine R, Barth J (2001) Bank regulation and supervision: what works best? The World Bank
Mahoney L, Roberts RW (2007) Corporate social performance, financial performance and institutional ownership in Canadian firms. Accounting Forum 31(3):233–253. https://doi.org/10.1016/j.accfor.2007.05.001
McDonald LM, Lai CH (2011) Impact of corporate social responsibility initiatives on Taiwanese banking customers. Int J Bank Mark 29(1):50–63. https://doi.org/10.1108/02652321111101374
McWilliams A, Siegel DS, Wright PM (2006) Corporate social responsibility: strategic implications 43(1):1–18. https://doi.org/10.1111/j.1467-6486.2006.00580.x
Melo T, Garrido-Morgado A (2012) Corporate reputation: a combination of social responsibility and industry. Corp Soc Responsib Environ Manag 19(1):11–31. https://doi.org/10.1002/csr.260
Mention AL, Bontis N (2013) Intellectual capital and performance within the banking sector of Luxembourg and Belgium. J Intellect Cap 14(2):286–309. https://doi.org/10.1108/14691931311323896
Michelon G, Boesso G, Kumar K (2013) Examining the link between strategic corporate social responsibility and company performance: an analysis of the best corporate citizens. Corp Soc Responsib Environ Manag 20(2):81–94
Miralles-Quirós MM, Miralles-Quirós JL, Redondo-Hernández J (2019) The impact of environmental, social, and governance performance on stock prices: evidence from the banking industry. Corp Soc Responsib Environ Manag 26(6):1446–1456. https://doi.org/10.1002/csr.1759
Montiel I (2008) Corporate social responsibility and corporate sustainability: separate pasts, common future. Organ Environ 21(3). https://doi.org/10.1177/1086026608321329

Morsing M, Schultz M (2006) Corporate social responsibility communication: stakeholder information, response and involvement strategies. Bus Soc Rev 15(4):323–338. https://doi.org/10.1111/j.1467-8608.2006.00460.x

Muniandy JK, Barnes L (2010) The link between corporate social performance and institutional investor's shareholdings in Malaysian public listed companies. Int J Bus Res Papers 6(4):246–261

Nandy M, Lodh S (2012) Do banks value the eco-friendliness of firms in their corporate lending decision? Some empirical evidence. Int Rev Financial Anal 25:83–93. https://doi.org/10.1016/j.irfa.2012.06.008

Oh W, Park S (2015) The relationship between corporate social responsibility and corporate financial performance in Korea. Emerg Mark Finance Trade 51:85–94

Oikonomou I, Brooks C, Pavelin S (2012) The impact of corporate social performance on financial risk and utility: a longitudinal analysis. Financ Manage 41(2):483–515. https://doi.org/10.1111/j.1755-053x.2012.01190.x

Owen D (2008) Chronicles of wasted time? A personal reflection on the current state of, and future prospects for, social and environmental accounting research. Account Audit Account J 21(2):240–267

Paulet E, Parnaudeau M, Relano F (2015) Banking with ethics: strategic moves and structural changes of the banking industry in the aftermath of the subprime mortgage crisis. J Bus Soc Rev 131:199–207. https://doi.org/10.1007/s10551-014-2274-9

Pérez A, Del Bosque IR (2013) Measuring CSR image: three studies to develop and to validate a reliable measurement tool. J Bus Soc Rev 118:265–286. https://doi.org/10.1007/s10551-012-1588-8

Perrini F (2005) Building a European portrait of corporate social responsibility reporting. Eur Manag J 23(6):611–627. https://doi.org/10.1016/j.emj.2005.10.008

Pistoni A, Songini L, Bavagnoli F (2018) Integrated reporting quality: an empirical analysis. Corp Soc Responsib Environ Manag 25(4):489–507. https://doi.org/10.1002/csr.1474

Porter ME, Van der Linde C (1995) Toward a new conception of the environment-competitiveness relationship. J Econ Perspect 9(4):97–118. https://doi.org/10.1257/jep.9.4.97

Rajan RG, Zingales L (2003) The great reversals: the politics of financial development in the twentieth century. J Financ Econ 69(1):5–50. https://doi.org/10.1016/s0304-405x(03)00125-9

Renneboog L, Ter Horst J, Zhang C (2008) Socially responsible investments: institutional aspects, performance, and investor behaviour. J Bank Financ 32(9):1723–1742. https://doi.org/10.1016/j.jbankfin.2007.12.039

Rowley T, Berman S (2000) A bra 25($nd new brand of corporate social performance. Bus Soc 39(4):397–412. https://doi.org/10.1177/000765030003900404

Rupp DE, Shao R, Skarlicki DP, Paddock EL, Kim TY, Nadisic T (2018) Corporate social responsibility and employee engagement: the moderating role of CSR-specific relative autonomy and individualism. J Organ Behav 39(5):559–579. https://doi.org/10.1002/job.2282

Saengchan S (2008) The role of intellectual capital in creating value in the banking industry. Int J Bus Res 7(2):157–169

Saravanamuthu K (2004) What is measured counts: harmonized corporate reporting and sustainable economic development. Crit Perspect Account 15(3):295–302. https://doi.org/10.1016/s1045-2354(03)00063-7

Saunders A, Allen L (2002) Credit risk measurement-new approaches to value at risk and other paradigms. Wiley, New York

Scholtens B (2009) Corporate social responsibility in the international banking industry. J Bus Ethics 86(2):159–175

Scholtens B, Dam L (2007) Banking on the equator. Are banks that adopted the equator principles different from non-adopters? World Dev 35(8):1307–1328. https://doi.org/10.1016/j.worlddev.2006.10.013

Schumpeter JA (1934) The theory of economic development. Harvard: Economic Studies

Scott JT, Pascoe G (1984) Capital costs and profitability. Int J Ind Organ 2(3):217–233. https://doi.org/10.1016/0167-7187(84)90011-0

Sethi SP (2005) Investing in socially responsible companies is a must for public pension funds—because there is no better alternative. J Bus Soc Rev 56:99–129. https://doi.org/10.1007/s10551-004-5455-0

Sharfman MP, Fernando CS (2008) Environmental risk management and the cost of capital. Strateg Manag J 29(6):569–592. https://doi.org/10.1002/smj.678

Shen CH, Wu MW, Chen TH, Fang H (2016) To engage or not to engage in corporate social responsibility: empirical evidence from global banking sector. Econ Model 55:207–225. https://doi.org/10.1016/j.econmod.2016.02.007

Shin MS, Rhee KN, Lee ET, Jung GJ (2018) Performance evaluation of CO_2 based ventilation control to reduce CO_2 concentration and condensation risk in residential buildings. Build Environ 142:451–463. https://doi.org/10.1016/j.buildenv.2018.06.042

Shleifer A, Vishny RW (1997) A survey of corporate governance. J Finance 52(2):737–783. https://doi.org/10.1111/j.1540-6261.1997.tb04820.x

Simpson WG, Kohers T (2002) The link between corporate social and financial performance: evidence from the banking industry. J Bus Soc Rev 35(2):97–109. https://doi.org/10.1023/a:1013082525900

Soana MG (2011) The relationship between corporate social performance and corporate financial performance in the banking sector. J Bus Soc Rev 104(1):133–148. https://doi.org/10.1007/s10551-011-0894-x

Statista (2020) www.statista.com
Statman M (2000) Socially responsible mutual funds (corrected). Financial Anal J 56(3):30–39. https://doi.org/10.2469/faj.v56.n3.2358
Thompson P, Cowton CJ (2004) Bringing the environment into bank lending: implications for environmental reporting. Br Account Rev 36(2):197–218. https://doi.org/10.1016/j.bar.2003.11.005
Valls Martínez MdC, Cruz Rambaud S, Parra Oller IM (2020) Sustainable and conventional banking in Europe. PLoS ONE 15(2). https://doi.org/10.1371/journal.pone.0229420
Valentine S, Fleischman G (2008) Ethics programs, perceived corporate social responsibility and job satisfaction. J Bus Soc Rev 77:59–172. https://doi.org/10.1007/s10551-006-9306-z
Venturelli A, Cosma S, Leopizzi R (2018) Stakeholder engagement: an evaluation of European banks. Corp Soc Responsib Environ Manag 25(4):690–703. https://doi.org/10.1002/csr.1486
Vurro C, Perrini F (2011) Making the most of corporate social responsibility reporting: disclosure structure and its impact on performance. Corp Gov 11(4):459–474. https://doi.org/10.1108/14720701111159280
Wahba H (2010) How do institutional shareholders manipulate corporate environmental strategy to protect their equity value? A study of the adoption of ISO 14001 by Egyptian firms. Bus Strategy Environ 19(8):495–511. https://doi.org/10.1002/bse.654
Weber O (2012). Environmental credit risk management in banks and financial service institutions. Bus Strategy Environ 21(4):248–263. https://doi.org/10.1002/bse.737
Windsor D (2006) Corporate social responsibility: three key approaches. J Manage Studies 43(1):93–114. https://doi.org/10.1111/j.1467-6486.2006.00584.x
World Bank (2020) Financial inclusion, Europe and central Asia economic update, Spring 2019
Wu MW, Shen CH (2013) Corporate social responsibility in the banking industry: motives and financial performance. J Bank Financ 37(9):3529–3547. https://doi.org/10.1016/j.jbankfin.2013.04.023
Wu J, Lodorfos G, Dean A, Gioulmpaxiotis G (2017) The market performance of socially responsible investment during periods of the economic cycle—illustrated using the case of FTSE. Manag Decis Econ. https://doi.org/10.1002/mde.2772

CHAPTER 4

Crowdfunding and Socially Responsible Initiatives

Lucia Michela Daniele

4.1 Fintech and the Alternative Financial Landscape: An Introduction to Crowdfunding

The Internet, and to an even greater extent digital technologies, represents a strong factor for discontinuity in the relationship between financial institutions and consumers. Created in response to the most recent financial crisis, alternative financial models rapidly took off worldwide as innovative ways of financing both businesses and social projects. The digitisation of financial services, coupled with greater access to information technology and wider use of mobile devices, has created fertile conditions for a new, disruptive, generation of firms that are redefining the financial value chain (Puschmann 2017). The fintech ecosystem is recognised as being one of the most important innovations of concern to the financial industry, given its profound effects on capital markets (Lee and Shin 2018). According to KPMG (2019), global investment in fintech startups in 2019 amounted to $135 billion from venture capital, merger and acquisition and private equity deals.

The Basel Committee on Banking Supervision (2018, p. 8) defines fintech as "technologically enabled financial innovation that could result in new business models, applications, processes, or products with an associated material effect on financial markets and institutions and the

F. Gangi et al., *The Evolution of Sustainable Investments and Finance*,
https://doi.org/10.1007/978-3-030-70350-9_4

provision of financial services". This umbrella term encompasses not only the provision of financial services through digital technologies but also new consumer behaviour and expectations, as well as the array of technological innovations (e.g. blockchain) that are likely to have a significant effect on financial markets and actors. As such, we can see that this is a highly heterogeneous industry in terms of players and segments served. Following Lee and Shin (2018), at the centre of this ecosystem are the fintech startups, businesses that rely on IT to unbundle[1] financial services, offering consumers access to numerous highly personalised and flexible financial services. Crowdfunding (CF) is the principal segment covered by fintech startups, on account of its ability to serve both individuals and businesses in the collection of alternative financial resources using digital platforms. The most recent data (CCAF, 2020) indicate that CF solutions represented nearly 84% of the total alternative finance market in 2018.

CF can be defined as "an open call, essentially through the Internet, for the provision of financial resources either in form of donation or in exchange for some form of reward and/or voting rights in order to support initiatives for specific purposes" (Belleflamme et al. 2014, p. 588), or as "the efforts by entrepreneurial individuals and groups – cultural, social and for-profit – to fund their ventures by drawing on relatively small contributions from a relatively large number of individuals using the internet, without standard financial intermediaries" (Mollick 2014, p. 2). The CF phenomenon has its roots in the concept of crowdsourcing, a term coined by Howe (2006) to describe the act of outsourcing to a large pool of external actors—such as online communities—tasks generally performed internally by organisations, in the form of an open call, and providing novel forms of compensation (Estellés-Arolas and González-Ladrón-De-Guevara 2012). Seminal contributions concur in defining CF as a form of crowdsourcing, through which individuals and organisations can present their projects via digital platforms to a large number of individuals potentially interested in supporting them financially (De Buysere et al. 2012; Schwienbacher and Larralde 2010; Belleflamme et al. 2014;

[1] The term "unbundling" refers to the segmentation of financial services between different operators—in other words, verticalised into specific financial segments, such as investments, payments, and loans to both businesses and individuals. While previously all these financial services were provided as part of integrated solutions offered by the incumbents, the development of digital technologies has led to a radical fragmentation of operators on the offer side.

Mollick 2014). Whereas in crowdsourcing, the "crowd" contributes with a variety of resources (knowledge, experience, etc.), in CF individuals supposedly contribute primarily with finance. CF therefore represents an alternative financial model to the extent that it enables individuals and organisations to launch an open call to the public, via a digital platform, with the aim of collecting financial resources from a large audience of both professional and non-professional investors.

The development of CF at the global level is linked to the growing relevance of sharing and collaborative models, in which the general public is invited to collaborate with other market players in value co-creation (VCC) processes (Prahalad and Ramaswamy 2004a, 2004b; Balboni et al. 2016). There are at least three main actors in the CF mechanism: the capital seeker or the CF proponent; the crowd; and the intermediary, namely the crowdfunding platform (CFP). Capital seekers are markedly heterogeneous, given that the crowdfunding process can be used for a wide range of purposes, such as philanthropic, creative, cultural, social innovation and entrepreneurial projects (Mollick 2014; Belleflamme et al. 2014). According to Hemer (2011), CF capital seekers might be independent entrepreneurs, integrated creators related to organisations (e.g. NGOs, foundations) or startups (e.g. entrepreneurial teams, new ventures). They start the CF process by storytelling their projects with the support of video and graphic materials on CFPs and employ a set of digital marketing strategies to attract the attention of potential supporters.

The crowd is the pivotal actor of the CF mechanism, being called on to judge the viability of the project proposed on the CFP, ultimately deciding whether to supporting it. The CF mechanism is often associated with the idea of the democratisation of finance and ideas, and also with the term "the wisdom of the crowd", which refers to the collective intelligence of a large group of people being able to deliver better decision-making than a restricted pool of investors (Surowiecki 2004). However, as discussed by Méric (2016), a crowd can be impulsive or suggestible, and its behaviour can be influenced by the marketing techniques of the CF experts. In line with Estellés-Arolas and González-Ladrón-De-Guevara (2012), and recognising that CF is a form of crowdsourcing, we view the crowd as a large group of individuals, the heterogeneity of which will depend on the type of initiative being funded. In fact, alongside segmentation according to type (e.g. professional or non-professional investors, interest in the project or cause, required expertise), we might find segmentation according to geographical location, for example, or connection with the project creator. Crowdfunders can be a close circle of the project creator's family and friends, or they can be anonymous and geographically distant (Mollick and Kuppuswamy 2014; Agrawal et al. 2015; Mochkabadi and Volkmann 2020).

Finally, Mollick's definition also highlights one of the main features of CF: the absence of intermediation by traditional operators. CFPs permit capital supply and demand to be matched, by engaging entrepreneurs and crowdfunders in digital spaces where they can exchange resources. CFPs create a multi-side market (Cumming and Hornuf 2018); they are virtual spaces relying on digital marketing logics to spread CF proposals to communities that are willing to invest. As well as facilitating capital exchange between capital seeker and crowd, these platforms offer technological infrastructures and legal and advertising services; their revenue comes principally from fees applied to the capital collected by the CF proponent (Agrawal et al. 2011; Moritz and Block 2016). As such, we can see the complexity of the CFP's role: it is not merely an intermediary but an environment in which communication and financial flows intertwine to and from different types of actor.

A number of prior studies have recognised that, beyond simply collecting financial resources, CF can confer additional strategic benefits such as a marketing function; pre-launches and market tests for new products and services; creating awareness of new initiatives; obtaining feedback and suggestions from the crowd; gaining visibility in professional investors' channels; and learning and entrepreneurial growth (Mollick 2014; Belleflamme et al. 2014; Wald et al. 2019). These features make the CF mechanism a highly flexible alternative financial model that can serve a number of different purposes (e.g. donations, pre-purchases of new products/services or equity and debit capital), as well as serving both individuals and different types of organisations (teams, new ventures, social enterprises, nonprofit organisations, civic communities).

4.1.1 Community and Financially Oriented Crowdfunding Models

From its inception, the development of CF has mainly been of interest to the cultural, philanthropic and creative industries. However, the spread of sharing platforms, linked to the credit crunch in the aftermath of the last financial crisis, has meant that CF has become a complementary source of capital for entrepreneurial initiatives and businesses (Cumming and Hornuf 2018; Block et al. 2018). The literature recognises four main types of CF models, and these can be further classified in two macro-categories: *community* and *financial*, according to the forms of compensation and crowdfunders' expectations (Kirby and Worner 2014). The "community" category encompasses all those CF models that offer emotional or material (non-financial) forms of compensation in exchange for financial support, such as donation-based or reward-based CF. The "financial" category includes all those models in which crowdfunders can invest capital with the aim of obtaining a financial return (Cumming and Hornuf 2018; Bottiglia and Pichler 2016).

In donation-based CF, the proposals are mainly linked to the third sector: a nonprofit organisation or a social enterprise might turn to the crowd to request financial support for a social, philanthropic or cultural cause. The funders in this model are mainly driven by philanthropic and social motives, rather than an expectation of some form of compensation. Reward-based CF is a model widely employed within creative and startup initiatives, as it allows crowdfunders to be recompensed without committing a large amount of additional resources to the project. In this model, the rewards that the creatives, innovators and entrepreneurs offer in exchange for the financial support are never of a financial nature; however, the crowd in this case are on the whole highly appreciative of small expressions of gratitude such as emails, public acknowledgement and personalised rewards (e.g. art, music, experience, new designs or innovative products) (Gerber and Hui 2016). The pre-purchase model is a growing sub-segment of the reward-based model, within which entrepreneurs are choosing co-investors in order to launch a new product or service (Ordanini et al. 2011). Here, entrepreneurs accept financial resources from the crowd in exchange for the pre-purchase of the product for which the capital is being collected. Among the community-based CF models, a highly specific form is civic CF, in which local communities and local governments collaborate to co-create public–private assets aimed at promoting and sustaining community social and environmental well-being (Davis 2015). In community-based models in general, crowdfunders (also referred to as supporters, backers or advocates) are motivated by the opportunity to support social and civic causes, to help entrepreneurs and organisations to bring innovative solutions or products to market, to be part of a community or to obtain exclusive and emotional rewards (Cholakova and Clarysse 2015; Gerber and Hui 2016; Bretschneidera and Leimeisterab 2017).

CF financing—or "crowdinvesting"—models cover a larger market share at a global level than community models, encompassing mature segments such as equity and lending CF, as well as fast-growing segments such as real estate CF (EY 2019; CCAF 2020). As Ahlers et al. (2015) explain, equity CF enables new firms to sell equity shares to small investors through CFPs. Investors purchase equity in the venture with the expectation of financial compensation in the form of future dividends or capital gains (Ahlers et al. 2015; Vismara 2016). There is another model within financing CF which in fact accounts for the majority of total CF market volume; this is the CF lending model, also referred to as "crowdlending". This model allows individuals to borrow money from other individuals or organisations with the expectation of profiting from interest and capital repayments at higher rates than those offered by the incumbents (Bottiglia and Pichler 2016). The rapid development of this model has given rise to two further sub-segments: P2P (peer-to-peer) lending, in which individuals exchange financial resources; and P2B (peer-to-business) lending, in which individuals invest debt capital in either startups or existing firms (Bottiglia and Pichler 2016). Finally, in recent years, spurred by the collapse of the real estate market following the 2008 financial crisis, the CF mechanism has been also been extended to investments in commercial and residential properties. Real estate CF allows individuals to invest both debt and equity capital in real estate projects, starting from relatively small amounts and offering them the opportunity to diversify their investment portfolio (Battisti et al. 2020).

4.1.2 *Signals in Crowdfunding: The Role of the Crowd*

The literature on the role of the crowd has mainly concentrated on the motivations behind supporters' participation in the CF process and on the factors that influence investment decisions (Gerber and Hui 2016; Moritz and Block 2016). Some studies have found support for the "wisdom of the crowd" argument by comparing, in terms of expertise, the selection of projects by crowdfunders with those of professional investors, concluding that the crowd is able to discern between "bad and good lemons" in the CF market (Mollick and Nanda 2016; Kim and Viswanathan 2018). A relevant stream of research in this context concentrates on signals produced by CF actors that reduce information asymmetry between crowd and proponent (Colombo et al. 2015; Cordova et al. 2015; Li and

Duan 2016; Courtney et al. 2017; Kim and Viswanathan 2016; Frydrych et al. 2016; Gangi and Daniele 2017).

Although CF is commonly considered a more flexible source of capital, in practice the achievement of financing objectives in this way is not so straightforward. It is not sufficient for entrepreneurs simply to launch a call to the crowd. Rather, from the entrepreneur's perspective, CF is a more democratic process and also a more onerous one as it involves considerable planning and communication efforts to reach and engage its audience. Information asymmetry is a notorious problem in the field of investments and relates to the disparity between how investors and capital seekers perceive the value of an investment, which affects decisions about both the cost of capital and the level of capital invested. Therefore, when an entrepreneur launches a CF campaign, he/she must consider the information and observable behaviours that the crowd might resort to in order to try to bridge the information gap and acquire the expertise needed to evaluate the CF proposal. When information asymmetries are not properly addressed, the risk of market failure increases, which, in terms of the CF activity, means failure to raise funds.

Information asymmetries demonstrate their own peculiarities in the context of a CF mechanism. First, the virtual environment of CF increases the risk of fraud: fictitious CF proponent profiles can be created, whose reliability might be not easily verifiable remotely (Tomboc 2013). Also, in contrast to more conventional forms of financing, in a virtual environment the investor cannot rely on "soft" information derived from personal interaction (Stein 2002; Petersen 2004)—or for that matter "hard" information from financial statements, considering that CF proponents are often teams or startups lacking verifiable track records. In reward-based CF—and the pre-purchase segment in particular—the relationship between the proponent and the crowd will not necessarily be a pre-contractual one. In addition, in reward-based CF there is no due diligence, and no governance mechanisms among CF entrepreneurs and CF co-investors. So, paradoxically, the theoretical democratisation of access to finance offered by CF might be derailed by information asymmetries between CF proponent and crowd.

Therefore, a critical component for success in a CF context is the leveraging by the CF proponent of signals that—directly or indirectly—increase the reliability and credibility of the CF proposal for the crowd. Following Spence (1973, p. 1), signals are those "activities or attributes which, by design or accident, alter the beliefs of or convey information to other individuals in the market". Their quality will depend on an unobservable ability of the signaller to fulfil the needs or demands of the outsider observing the signal (Connelly et al. 2011).

The CF fundraising time frame can be divided into three main phases (Kuppuswamy and Bayus 2018; Beaulieu et al. 2015): starting, middle and final. Generally, the starting and final phases are characterised by high supporter participation, but the middle phase often demonstrates a decline in contributions. The point about signals is relevant here. Each phase is usually characterised by a different set of signals produced by the various actors, whether simultaneous or not, that contribute to the creation of a collective judgement about the CF proposal. During the starting phase, signals are generally relayed by the CF proponent and early investors (Mollick 2014; Frydrych et al. 2014, 2016; Koch and Siering 2015), whereas the middle and final phases are characterised by the information that entrepreneurs send to the crowd (e.g. updates, new stretch goals or rewards) and by the information produced by the first wave of late supporters, CF ambassadors and virtual communities via word-of-mouth mechanisms.

Prior investigations show that, because communication about the characteristics and purpose of a CF project occurs in a virtual context, high-quality signals about the preparation and launch of a campaign are positively associated with success (Mollick 2014; Frydrych et al. 2014; Koch and Siering 2015; Courtney et al. 2017). Following the literature (Mollick 2014; Frydrych et al. 2016; Gangi and Daniele 2017), success appears to be linked to project quality and presentation in terms of accuracy of project description, the presence of video material, updates and plenty of P2P links on social networks. The CF mechanism by its very nature presumes high involvement with the wider public, so network dynamics and signalling power are highly relevant. Other investigations, from a social capital theory standpoint, highlight that social networks, social media and a proponent's individual social capital are also drivers of success in the CF market (Hardy 2013; Cordova et al. 2015; Colombo et al. 2015; Balboni et al. 2016). Several studies confirm that social networks (e.g. Facebook, Twitter, LinkedIn) and the "e-reputation" of a CF proponent, as substantiated by his/her personal and professional networks, reinforce a CF proposal's legitimacy and thus its success. In addition, a proponent's prior CF reputation is seen as a success driver, with several studies demonstrating that entrepreneurs that frequently have recourse to CF receive higher levels of finance than their competitors

(Skirnevskiy et al. 2017; Butticè et al. 2017). Further, CF engagement in terms of support for other campaigns and entrepreneurs also signals a CF proponent's credibility to the crowd. The CF proponent acquires a "digital reputation" and increases his/her legitimacy among the CFP communities. Reputational signals are also transmitted when the CF proponent or team are seen to have already developed skills and experience within the sector of the CF proposal.

As well as the signals sent by the entrepreneurs, there are those generated by the crowd, which, from a social information theory perspective, are important in the creation of a collective judgement about a CF proposal. Social information theory affords two perspectives in explaining crowd behaviour in a CF context. First, in line with the diffusion of responsibility model (Fisher 2011), the probability that an individual will support an initiative decreases according to the perceived or actual presence of other people who have already decided to support it. In essence, the model assumes that people avoid being engaged in initiatives if they believe that others are already doing it. This framework is useful in interpreting crowd behaviour in the middle CF phase, where the most promising CF campaigns may have already reached their financial target and further potential supporters might fail to be convinced, and await further signals (Kuppuswamy and Bayus 2018). Nonetheless, the diffusion of responsibility model fails to explain crowd behaviour in the start and end phases.

Turning to other studies (Belleflamme et al. 2014; Cordova et al. 2015), a second mechanism for interpretation of crowd behaviour is the so-called reinforcement effect or model, according to which agents or investors follow what they perceive others to be doing. In fact, the crowd in a CF context see other agents' actions as a form of observational learning source, the better to assess the quality of a project that is not directly observable. Indeed, the evidence in prior studies converges to support the idea of herd behaviour among the crowd. As Cordova et al. (2015) reveals, supporters in a CF mechanism see the number of contributions received by a project as a signal of quality. Advocates are more likely to finance projects that seem to have a greater chance of success and, conversely, shun those that are far off their targets. So, from a herd behaviour perspective (Belleflamme et al. 2014; Kuppuswamy and Bayus 2018), the number of backers is correlated with success. Similarly, the signals sent by different categories of supporters, such as early or late backers, to subsequent supporters vary in quality.

Observing reward-based crowdfunding, Colombo et al. (2015) empirically demonstrate that early contributions positively affect further contributions in subsequent phases of the campaign. Gangi and Daniele (2017) expand the investigation into the role of behavioural influence among backers, showing that supporters coming in at the end of the campaign also boost financial receipts. Further signals can be transmitted to the crowd by the ambassadors, individuals and organisations that act as experts or mentors for initiatives, and these also increase the chances of success (Mollick 2016; Gangi and Daniele 2017). Expert CF investors or supporters generally intervene in the early stages of a campaign, by choosing the best reward–price combination. The behaviour of these supporters tends to positively influence the non-experts either with respect to the product and/or service being offered or with respect to CF mechanisms more generally. The co-financing of a CF campaign by an organisation (so-called third-party endorsement) or a big company is another reliable signal about the credibility of a CF proposal (Gangi and Daniele 2017), mitigating information asymmetries in the process and thus increasing the chance of success. Expert reviewers represent another relevant source of information in evaluating a product. Endorsements from informed third parties, such as product reviews by a notable online exponent in the sector, may offer information about the project that encourages potential supporters (Chen and Xie 2005; Zhu and Zhang 2010).

Kuppuswamy and Bayus (2018) point out that crowd orientation towards herd behaviour, or the diffusion of responsibility model, is moderated by the deadline effect. Due to the tendency of individuals to procrastinate with regard to both decisions and actions (Ariely and Simonson 2003), the deadline effect offers an explanation of why many CF contributions occur in the closing phase. When a deadline is looming, people are more likely to act, even if they see others already offering support; at that point potential backers have lower expectations of how much support a project will ultimately achieve overall, which also spurs supporters to contribute near the project's deadline. This accords with Kuppuswamy and Bayus (2018) finding that successful projects see much of their support in both the early and the last weeks of the funding cycle; the deadline effect means backers' support is negatively correlated with previous backers, but only up to the moment at which the deadline effect diminishes as the project approaches its closure time.

Finally, feedback from the crowd represents a helpful additional source of information. CFPs generally provide digital interfaces via which CF proponents can enter into dialogue directly with supporters through the comments functions. CF marketing activities are also a critical asset through which to engage supporters and investors by means of social networks. Crowd insights and feedback can help make the CF proposal conform to market expectations, and furthermore augment the financial and knowledge resources acquired by the CF proponent (Skirnevskiy et al. 2017). The proponent can therefore make use of early supporters' (direct or indirect) input to modify or upgrade the project in anticipation of negative scenarios or to increase the chances of engaging with and meeting potential supporters' expectations. As such, a larger number of supporters means a greater opportunity to fine-tune the project. Those projects with numerous early backers are more likely to become consolidated in a form that meets the general approval of the crowd, thus increasing the probability of the campaign reaching its target (Colombo et al. 2015).

This informational cascade among the different crowdfunding phases (Thies and Wessel 2014) is favoured by the word-of-mouth (WOM) process, which, as in online auctions (Ariely and Simonson 2003), plays a role in the CF mechanism. The WOM generated around a campaign crystallises the collective judgement about a crowdfunding proposal, which, via a CFP, is formed by and made available to a multitude of individuals and organisations. Supporters converge in online communities via CFPs, where social information and social networking activities allow a collective culture to emerge—the wisdom of the crowd—from which individuals can acquire information from outside their own competences and experience. Traditional (offline) WOM has been shown to play a major role in customers' purchasing decisions (Hennig Thurau et al. 2004). In a virtual environment, too, reviews by experts or influencers, comments, opinions and ratings have been proved highly influential in shaping consumers' behaviour and perceptions about a product (Rindova et al. 2007). Following Bi et al. (2017), an online consumer has two routes of information on which to rely: central and peripheral. The peripheral route concerns all the information coming from the environment in which the product is embedded, such as reviews by other consumers in the online stores (i.e. WOM). The central route represents information provided by the entrepreneurs themselves. Bi et al. (2017) find that, in a CF context, both the central and peripheral routes have a positive effect on the likelihood of financial success. WOM positively affects CF success further by augmenting the level of the agent's information about proposed projects and initiatives.

4.2 The Relevance of Crowdfunding in Entrepreneurial Finance

Because entrepreneurs can collect funds from geographically distant and non-expert investors, instead of relying on a restricted pool of professionals, researchers have referred to CF as one of the most significant developments in the democratisation of finance (Mollick and Nanda 2016; Stevenson et al. 2019; Cumming et al. 2019). It refers to the opportunities generated by information and communications technology (ICT) developments to involve the general public in the early-stage funding of new ventures, thus lowering the information barrier between entrepreneurs, particularly under-represented ones, and the capital markets. CF is generally considered a more transparent and democratic way for entrepreneurs to collect financial resources (Méric 2016; Stevenson et al. 2019), as it involves the intervention of a multitude of individuals, whose heterogeneity potentially mitigates any bias associated with the incumbent's screening processes, thus making access to financial resources more democratic (Cumming et al. 2019; Ramos 2018).

The problems that have always attended the sourcing of external finance for new initiatives at reasonable rates (e.g. information asymmetries, opacity and risk, or gender- or race-related biases) (Akerlof 1970; Cumming et al. 2019) were exacerbated by the 2008 financial crisis (Block et al. 2018; Cumming and Hornuf 2018). The credit crunch impaired banks' ability to lend capital to businesses, which was also affected by more stringent patrimonial regulations introduced by the Basel Committee to prevent further financial turbulence. Concurrently, the emergence of platforms enabling peer-to-peer resource exchanges allowed organisations to search for alternative ways to seek finance (Block et al. 2018; Cumming et al. 2019). As a result, digital technologies have had a huge impact on the way new and smaller companies are financed, and CF has been at the forefront in bridging the gap between supply and demand of entrepreneurial finance in both developed and less developed economies (Mollick 2014; Bruton et al. 2014; Cumming and Hornuf 2018). This is reflected in the latest data available on the crowd financing industry and the growing use of alternative financing instruments by small and medium enterprises (EY 2019; OECD 2019; CCAF 2020) (Fig. 4.1).

According to CCAF (2020), in the CF market, equity and lending models predominate over non-investment-based CF models. P2P lending CFPs intermediate between potential lenders willing to invest to obtain a return from alternative investments and borrowers that can be either individuals or businesses (Bottiglia and Pichler 2016). In a lending CF, while investors are motivated by the opportunity to place their resources among a diversity of different investment options, borrowers are looking for an alternative capital-raising channel to the traditional ones where they might encounter obstacles (Iyer et al. 2009; Bottiglia and Pichler 2016). Emerging as a personal finance tool for the exchange of loans between individuals, the lending CF mechanism rapidly evolved into a model offering solutions for businesses. Recent data (CCAF 2020) on lending CF for businesses reveal a lower onboarding rate but a higher success rate compared to the consumer segment. Accordingly, from the capital debts side, lending CF is moving towards becoming a complementary source of capital for small and medium enterprises (Pierrakis 2019; Bottiglia and Pichler 2016). As far as Pierrakis (2019) is concerned, the rapid growth of this model can be attributed to a more sophisticated public of investors, quite different in terms of expertise and expectations

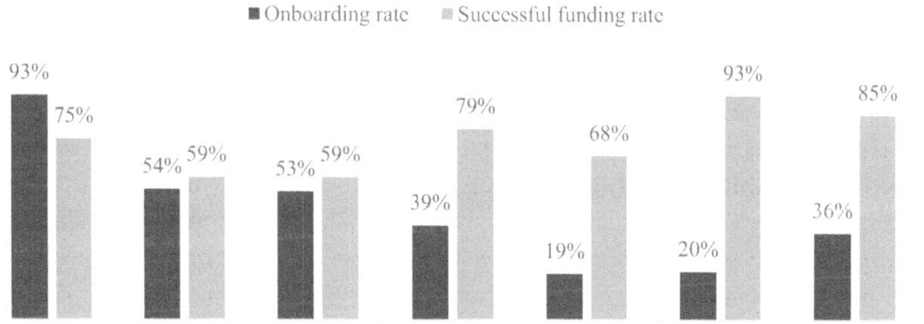

Fig. 4.1 Global onboarding and success rate of CF models (*Source* Authors' elaboration of CCAF data, April 2020, in *The Global Alternative Finance Market Benchmarking Report*. Onboarding rate: the rate at which CF proponents are allowed to proceed with a fundraising campaign. Successful funding: percentage of onboarded firms that effectively collected capital from the platform)

from investors operating in other CF models. Crowdlenders generally have investment experience with other financial solutions; variables influencing their decisions to lend money relate mainly to a specific firm's risk profile, while they are driven principally by expectations of making a financial return (Pierrakis 2019).

Unlike traditional early-stage financing, the main idea of equity CF is to raise capital by obtaining small contributions from a large number of investors. Equity CF has a lower onboarding rate than the lending segment for business (19% and 39%, respectively), implying that this model generally presents higher entry barriers. However, as the literature contends (Mochkabadi and Volkmann 2020; Cumming and Hornuf 2018; Vulkan et al. 2016), this model continues to represent a viable and efficient alternative source of capital for new ventures, as signalled by the successful funding rate of equity CF campaigns (68%). On account of its relevance to the alleviation of barriers to finance for new entrepreneurial initiatives, the configuration of information asymmetries has remained the chief focus of the literature on equity CF (Mochkabadi and Volkmann 2020; Cumming and Hornuf 2018). Research on the signals that entrepreneurs can send to the crowd in pursuit of credibility and success is related mainly to the following: level of detail and coherence of presentation (Ahlers et al. 2015; Vulkan et al. 2016); human and social capital (Piva and Rossi-Lamastra 2018; Nitani and Riding 2017; Vismara 2016); third-party endorsement and assets (Vismara 2018; Ralcheva and Roosenboom 2016); updates and high-quality communications (Angerer et al. 2017; Nevin et al. 2017); and founders' commitment (Shafi 2019). The funding level, as well as the portion of capital offered, also has a bearing on the success of equity CF campaigns (Ahlers et al. 2015). In reward-based CF, requests for higher levels of funding might discourage crowd participation (Vulkan et al. 2016; Löher et al. 2018). And the share of equity offered to investors implies the share of risk that entrepreneurs are asking the crowd to bear. The herd behaviour mechanism is relevant in the equity CF model as well, with the number of investors and the information cascades among investors leading to higher success rates (Vismara 2018; Vulkan et al. 2016; Hornuf and Schwienbacher 2018). Nitani et al. (2019) put forward a "wisdom of the crowd" argument with regard to CF equity, in contrast to Piva and Rossi-Lamastra (2018), by providing evidence that the crowd is able to interpret signals embodied in CF campaigns and proponents' networks to inform its investment decisions. Indeed, given the similarity between the signals that the crowd has at its disposal in choosing among CF equity investments and those that professional investors use (Vulkan et al. 2016; Mochkabadi and Volkmann 2020), some studies have concentrated on interactions occurring in CFP environments between the crowd and professional investors. Professional investors might use equity CFPs to diversify their investments, while their presence there as co-financers signals the provisional ability of a new venture to attract the interest of the general public (Cumming et al. 2018).

The various crowdinvesting models have attracted the interest of numerous researchers because of their ability to meet the financial needs of new ventures and SMEs. Although the reward-based CF is not strictly an investment-based model, there is empirical evidence of its ability to support new ventures, especially the pre-purchase segment (Mollick and Kuppuswamy 2014). As well as offering the opportunity to test the market appeal of a new proposal (Mollick and Kuppuswamy 2014; Belleflamme et al. 2014), CF is attractive to entrepreneurs as a means of launching ventures because the crowd validation mitigates adverse selection by professional investors, enabling a startup to obtain equity capital more easily in subsequent investment rounds (Drover et al. 2017). Entrepreneurs also attribute to the CF mechanism other positive externalities such as the opportunity to engage consumers in the vision of a new venture, to obtain feedback and useful insights from the crowd, to collect financial resources at a lower cost, to create awareness and visibility around a new business project or to activate a co-creation process with the crowd and learn about market preferences (Gerber and Hui 2013; Butticè et al. 2017). These positive externalities, deriving from interactions with different actors, have given rise to pre-purchase CF as a model that supports entrepreneurial finance, as demonstrated by the presence of serial entrepreneurs in the reward-based CF model. Such serial creators are those entrepreneurs who repeatedly use CF to grow their ventures by launching consecutive projects, exploiting the opportunity to attract a virtual community with which to develop both commercial and emotional connections (Butticè et al. 2017). Research has demonstrated how social capital accumulation through CFPs has led serial CF entrepreneurs to benefit from the establishment of long-term ties with communities, increasing the likelihood of turning supporters into loyal financial backers (Skirnevskiy et al. 2017; Butticè et al. 2017), in contrast to the experience of a CF first-timer. This finding is supported by Cholakova and Clarysse (2015), for whom crowdfunders' motivations in reward-based

CF campaigns are primarily finance- or utility-driven, rather than social. In line with this, studies on reward-based CF show that consumers are replacing the initial funding sources (Mollick and Kuppuswamy 2014; Skirnevskiy et al. 2017; Butticè et al. 2017), contrasting with prior evidence that suggested that the majority of CF contributions are from proponents' networks of family and friends (Agrawal et al. 2015).

The majority of studies currently available on CF refer to the impact of specific crowdinvesting models in relaxing the financial constraints for new ventures. It should be noted that CF models are not mutually exclusive, since entrepreneurs can collect financial resources by combining different forms of CF and can opt for different models of CF according to the entrepreneurial stage in which their businesses find themselves (Belleflamme et al. 2014; Paschen 2017). Earlier studies (Belleflamme et al. 2014; Miglo and Miglo 2019) analysed the economic convenience factors that might lead entrepreneurs to discriminate among equity and reward-based models to fund their venture. Pre-purchase approaches are appropriate for entrepreneurs seeking a relatively small amount of finance and who intend to present a new product or service aiming to reach a large base of early adopters (Belleflamme et al. 2014; Miglo and Miglo 2019). Paschen (2017) provides an exhaustive picture of which CF model should be considered most appropriate for each startup phase. Community-based CF models are especially indicated for the pre-startup phase, as in this case the risk associated with the initiative is high and the entrepreneur has neither the financial resources nor the operational track record to offer rewards or prospective financial compensation (Paschen 2017). As the business gains visibility and proceeds towards the startup phase, then an entrepreneur might consider a pre-purchase scheme to validate the product and establish a base of early adopters (Paschen 2017; Ordanini et al. 2011; Mollick 2014). Additionally, a lending CF can be a highly flexible and sustainable source of capital at this stage: lending CFPs offer the opportunity to take out relatively small loans on more flexible terms compared to banks (Bottiglia and Pichler 2016; Paschen 2017). Finally, the more mature stages allow the possibility of trying a profit-sharing model such as equity CF. By the time this stage has been reached, startups have already established a business model and the risk of project failure is lower than at the inception. Generally, ventures applying for equity CF are more transparent, have track records and are committed to improving their business model to scale for the market (Paschen 2017;

Miglo and Miglo 2019). Hence, here CF can enlarge funding occasions for new ventures, making their financial structure more flexible and providing them with different models with which to integrate funding strategies supporting diverse business goals.

4.3 The Crowdfunding Ecosystem for Innovation and Value Co-creation

The "wisdom of the crowd", as stated, generally refers to the result of a collective effort that is more effective than decisions made by a restricted pool of individuals: interactions and links among groups promote diversity, independence and knowledge aggregation which all make for better decision-making (Brabham 2008; Surowiecki 2005; Hong and Page 2004). The potential of this crowd wisdom, coupled with developments in IT as mechanisms for distributing knowledge and information, is as a form of distributed intelligence that leads businesses to engage in working out how to benefit from open innovation processes and crowdsourcing practices to increase their understanding of future market trends (Chu et al. 2019; Estellés-Arolas and González-Ladrón-de-Guevara 2012). In the CF literature, evidence abounds of how the social capital related to CF proponents and their networks, or related to the CF communities, drives the success of CF campaigns and ensures a series of positive externalities for organisations in terms of intangible resources (Mollick 2014; Colombo et al. 2015; Vismara 2016; Butticè et al. 2017; De Luca et al. 2019). Following Nahapiet and Ghoshal (1998, p. 243), social capital encompasses not only the network itself but also the "assets that may be mobilised through that network", and, as such, crowdsourcing practices in online settings, such as CF, align stakeholder networks in a configuration in which they can interact and generate additional value. In a broader view, CF is a process of assembly of resources from numerous actors who collaborate to support a specific purpose (Belleflamme et al. 2014; Mollick 2014). Accordingly, in a CF context, the social capital perspective can be observed both at individual and collective levels (Giudici and Rossi-Lamastra 2018), namely as an output of CF campaigns rather than simply a prerequisite for success.

Based on these observations, researchers have begun to look at the social capital collective dimension of the CF mechanism and related benefits in terms of innovation (Chu et al. 2019; Giudici and Rossi-Lamastra 2018; Stanko and Henard 2017; Mastrangelo et al. 2019; Azar and

Mackey 2015). In line with De Luca et al.'s review (2019), the innovative potential of CF might affect different strategic business areas (e.g. finance, marketing, CSR, production and design), because its large audience allows for a wider range of goals compared to the limited number of actors in a traditional financial channel. From a knowledge creation perspective, CF-related studies demonstrate that communities represent a reservoir of open-source knowledge from which ideas and feedback from external and highly heterogeneous entities can be tapped, creating conditions for radical innovation and higher product performance in the pre-purchase mechanisms (Stanko and Henard 2017; Chu et al. 2019). Azar and Mackey (2015), from a finance-for-innovation perspective, conclude that CF is a particularly viable option in the early phases of biotech startups' R&D development. Chu et al. (2019), analysing the characteristics of open innovation in the CF context, conclude that this model can be interpreted as an "intermediary space", where value co-creation (VCC) opportunities emerge from the interaction between CFPs, CF main actors and third parties. Similarly, investigating the potential of CF for SME innovation, Giudici and Rossi-Lamastra (2018) concur in recognising collective social capital and open approaches to innovation as drivers of credibility in equity CF, increasing the success of the assembly of financial resources.

Hence, CF allows the crowd to engage in different roles and forms of VCC with CFPs and CF proponents and allows organisations to leverage their innovative ability using assets and knowledge obtained from the crowd. Despite the social capital collective dimension being generally attributed to the whole CF phenomenon, VCC processes are easier to observe in the pre-purchase reward-based model, where the interactions with early adopters create opportunities for entrepreneurs to accrue crucial knowledge about their product's ability to fulfil actual market needs (Mastrangelo et al. 2019; Lipusch et al. 2020; Stanko and Henard 2017).

The VCC paradigm is a relatively new idea and refers to a multilateral, open and dialectical process of interaction and knowledge sharing among market entities, whereby the parties engage in a dialogue to jointly define reciprocal beneficial solutions (Ranjan and Read 2016; Prahalad and Ramaswamy 2004a). The fundamental premise of the VCC process is cooperation of market entities (Cova et al. 2011). Actors involved in a VCC process are viewed as resource integrators, who collectively create an ecosystem of shared value to perform collaborative activities and are encouraged by the possibility of achieving different kinds of benefits (Galvagno and Dalli 2014; Prahalad and Ramaswamy 2004b). Resources exchanged can vary from the tangible to the intangible; they may or may not be highly specialised and can be related to either a functional or an emotional sphere (Payne et al. 2008). With regard to VCC actors' expectations, Gupta and Lehman (2005) clarify that, since different actors are participating in the co-creation, all perspectives need to be taken in account in order to fully capitalise on VCC's benefits. Actors are attracted to a VCC process because of expectations of new experiences (Prahalad and Ramaswamy 2004a, 2004b), the value of which is "uniquely experienced and determined by the beneficiary" (Vargo et al. 2008). This helps to explain why, in a VCC context, such as a CF campaign, the reasons underpinning the virtual community's collaboration are related to the ability of proponents to set appropriate incentives, either monetary or more intrinsic or subjective.

In a review of 149 papers, Ranjan and Read (2016) investigate the constituencies underlying VCC processes, namely the co-production (CP) and the value-in-use (ViU) axioms and related sub-dimensions. CP domains are activities and resources exchanged by actors to co-create a new market proposal, whereas in the ViU concept value is generated by the interaction of the consumer with the market entity's proposals, and thus by the value that the former associates with the co-creation experience.

The CP domain encompasses knowledge sharing, equity and interaction. Knowledge sharing represents a fundamental condition, as the level of information disseminated among actors impacts the level of actionable co-creation processes (Ranjan and Read 2016; Chu et al. 2019). CFPs provide space where the proponents can co-create value through the exchange of information. Additionally, crowdfunders co-define the CF proposal by offering diversified competences and expertise (Kim and Viswanathan 2018; Mollick and Nanda 2016). This creates opportunities to expand collaborative opportunities, as crowdfunders can perform several roles simultaneously, from co-ideation and evaluation of an idea, to co-investment in and co-testing of a new market launch (Russo-Spena and Mele 2012; Ordanini et al. 2011). Hence, crowdfunders can choose whether, and with what level of intensity, to co-create with organisations, in accordance with the benefits expected from the VCC process. If a supporter finds a CF proposal attractive, he/she can choose to limit his/her commitment to a simple financial contribution, or be more proactive—socialising with other supporters, acting as an ambassador, or performing a more specialised activity, such as contributing feedback and insight to the product's design and test. The increased level of collaboration that supporters are offered also affects how they perceive their own responsibility in contributing to the final outcome (Ordanini et al. 2011; Hu et al. 2015). In line with Ranjan and Read's (2016) VCC framework, in the reward-based model the equity sub-dimension makes for a psychological ownership of CF outcomes, which in turn means increased willingness to support them to succeed. Finally, following Ranjan and Read (2016), interactions favour co-production, as social and informational ties enable actors to influence one another and signal their commitment to the success of the collaborative process. In CF studies, crowd herd behaviour theories help to explain how different informational flows (both within crowdfunder networks and from outside them) create patterns of reciprocal influence among crowdfunders, and between crowdfunders and other third parties, such as reviewers, influencers or mentors (Colombo et al. 2015; Zhu and Zhang 2010; Courtney et al. 2017; Gangi and Daniele 2017).

The second underlying dimension of VCC, ViU (Ranjan and Read 2016), encompasses experience, personalisation and relationship. From the perspective of Prahalad and Ramaswamy (2004a), who describe the market as a "forum for co-creation experiences", VCC processes enable users to attain unique value, as such practices allow them to calibrate their experience and their degree of engagement in accordance with their needs and expectations. With regard to the personalisation sub-dimension, in pre-purchase mechanisms the range of expectations about co-creation activities among the crowd and other CF actors includes both material and emotional benefits. Early adopters of a new product in a CF context are generally motivated by the possibility of being one of a privileged minority to obtain the product and/or exclusive materials and of helping to refine the design. The same process is at work in a reward-based CF, but here communitarian and social motivations are also relevant. Social recognition—belonging to a community of innovative consumers—as well as the opportunity to share knowledge with more informed and expert backers are all factors incentivising crowdfunders' involvement in VCC activities with organisations. ViU, as a benefit for the crowd, also resides in social and relational opportunities. The collective process of a CF campaign reinforces backers' personal values, such as cooperation and reciprocity (Ranjan and Read 2016; Chandler and Vargo

2011). The CF literature supports community relationships as a critical driver of crowdfunders motivation (Koch and Siering 2015; Gerber and Hui 2013).

Applying the VCC paradigm to the CF domain sheds light on how this model offers co-creation experiences for the CF actors. A CF campaign is an open call to the general public to support a project proposal, the financial success and future development of which will depend on the degree of cooperation among different parties. The main parties are the community of investors, the project proponent and the CFP. However, the success of the undertaking depends on all the resources provided and integrated by these actors, in terms of financial information, track record, visibility, marketing and networking. This is evident from the emerging stream of literature that finds support for VCC in CF as an innovative new way of overcoming information barriers and financial hurdles.

Identifying personal benefits for entrepreneurs accruing from CF, Frydrych and Bock (2018) cite the opportunity to co-create their identity, and subsequently their reputation, with the crowd in order to leverage their chances of success. Also looking at benefits for entrepreneurs, Mastrangelo et al. (2019) assert that VCC in a CF environment positively affects entrepreneurs' ability both to secure financial resources and to achieve personal success. These results are in line with Kim and Hall (2019), who empirically demonstrate that, in the tourism sector, entrepreneurs proactively co-creating with other CF stakeholders (e.g. investors, sites, customers, CFPs) leads to increased crowd participation in CF campaigns. Finally, Lipusch et al. (2020) focus on features of CFPs that enable co-creation opportunities for product development, revealing that crowd participation is mediated by co-creation opportunities and the feelings of psychological "ownership" associated with such opportunities.

In sum, crowdfunding is a novel value co-creation model that comes to life through crowdfunding platforms—virtual spaces that provide organisations with the technological, financial and social resources so they might benefit from the engagement of individuals and communities (Ramos 2018; Kim and Hall 2019; Chu et al. 2019; Quero and Ventura 2019; Mastrangelo et al. 2019; Lipusch et al. 2020; Ordanini et al. 2011; Colombo et al. 2015).

4.4 Crowdfunding and Sustainability

There are many factors behind the growing importance of sustainability, and these are deeply interconnected globally. They encompass climate change, increasing scarcity of natural and food resources, the complex dynamics of globalisation, such as migration, the related issue of the democratic integration of cultures (UN 2020; Gangi et al. 2019a) and risks associated with pandemics. These are some of future challenges underscoring the interest in the concepts of corporate sustainability (CS) and corporate social responsibility (CSR). CS and CSR are convergent concepts, sharing the same vision about the need to balance economic responsibilities with social and environmental responsibilities. Following van Marrevijk (2003), CSR concerns transparency and dialogue with stakeholders and social reporting, whereas CS focuses on the pillars that support a sustainable approach, such as value creation through environmental production processes and environmentally friendly behaviour, stakeholder governance mechanisms, and socially responsible human resources management.

From a stakeholder theory perspective (Freeman 1984; Dahlsrud 2008), an organisation committed to CSR is one that recognises that it is a part of society, not separate from it, and which is aware that organisational performance and long-term viability increasingly depend on a wide spectrum of relationships with its stakeholders (Andriof and Waddock 2002; Barnett and Salomon 2012; Tang et al. 2012). Sustainability has therefore become synonymous with opportunity, i.e. a new strategic perspective to address the complexity that characterises the environmental challenge, guaranteeing the company's survival and prosperity over the long term, due to a greater convergence between corporate results and collective expectations.

As discussed, CF represents a valuable tool for innovation, being not only an online fundraising activity but an innovative logic for connecting entities, communities and people. In a Schumpeterian perspective, the CF phenomenon is a social innovation mechanism that can induce sustainable outcomes from the collaboration of a network of multiple stakeholders, leading to what Gummesson (2006) defines as a centric balanced outcome. Representing a many-to-many network, this alternative financial model can contribute to the creation of long-term relationships and well-functioning markets by promoting VCC opportunities in accordance with multiple stakeholders' expectations; as such, it then has the potential for a huge impact on society as whole in terms of sustainable development (Agrawal et al. 2015). Studies already cited here suggest that the social innovation mechanism of CF could foster democracy and transparency, aggregating ideas and resources around valuable projects, and potentially contributing, via the investment process, to solutions to urgent environmental and social sustainability issues.

The discourse around CF and sustainability revolves mainly around three points: the synergies between CSR engagement and CF (Spanos 2018; Marom 2017; Ramos 2018; Behl and Dutta 2019); CF's potential to promote financial inclusion by serving financial minorities (Lehner 2013; Lehner and Nicholls 2014; Balboni et al. 2016; Cumming et al. 2019); and the potential to leverage CF success via a CSR orientation (Allison et al. 2015; Mastrangelo et al. 2019; Behl and Dutta 2019; Flórez-Parra et al. 2020; Berns et al. 2020; Wehnert et al. 2019).

4.4.1 Crowdfunding and CSR

According to the CSR–CFP research stream, organisations engaged in trustful relationships with stakeholders and expressing higher CSR commitment enjoy, as a consequence, better organisational performance and a better corporate reputation (Gangi et al. 2020; Surroca et al. 2010). A socially responsible company in elevating its reputation bolsters its credibility among its interlocutors, both actual and potential, and enjoys benefits otherwise not possible. Hence, corporate reputation (CR) and credibility are strategic assets for maintaining competitive advantage: organisations that approach CSR as an opportunity in which they benefit from stakeholder engagement accrue a greater stakeholder influence capacity (SIC) (Barnett 2007). Social reporting and communication have assumed an important role in CSR discourse as critical components of corporate reputation (Brammer and Pavelin 2006; Bhattacharya and Sen 2004). This intangible asset measures stakeholders' perceptions and expectations of the organisation's conduct, including its contributions to environmental protection, social capital and equality (Fombrun et al. 2000).

Technological development has led to a dual transformation in CSR engagement: the creation of both new spaces and new forms of dialogue between organisations and stakeholders (Illia et al. 2017). Digital spaces (e.g. social media, sharing platforms) allow organisations to directly monitor information flows to and from stakeholders, with significantly higher time- and cost-efficiency compared to traditional channels (Wanderley et al. 2008; Kesavan et al. 2013). This allows the collection of real-time perceptions from stakeholders about issues such as CSR strategies and reports (de Bakker and Hellsten 2013; Illia et al. 2017). Researchers have observed CSR engagement and CFP communication shifting progressively from institutional business sites to social networks and digital platforms, where actors can exchange real-time information and opinions (Illia et al. 2017; Grigore et al. 2017). Likewise, prior studies demonstrate that traditional non-financial reporting systems are better suited to dialogue with professional investors, whereas non-professional investors are more influenced by digital media (Axjonow et al. 2018).

4.4.1.1 CFPs' Socially Responsible Orientation

As explained above, a CF mechanism implies the collaboration of at least three parties: namely, the CF proponent, the crowd and the CFP. However, as CFPs are virtual spaces typically powered by social media communication, their viral networking activity allows ideas from different stakeholder networks to emerge (e.g. customers, suppliers, public institutions, professionals, the general public), leveraging the likelihood of CF success (Stanko and Henau 2016; De Luca et al. 2019; Frydrych and Bock 2018).

From the perspective of CF's main actors, such as the CFPs, the growing relevance of CSR engagement, and related ESG ratings, might be a critical factor in the long-term economic sustainability of their business model, as visibility and credibility in the market is a critical efficiency factor (Marom 2017; Ramos 2018). Indeed, as noted by prior studies (Marom 2017; Ramos 2018), these sites are businesses, and as such can be expected to pay attention to relevant issues in the financial sector, such as CSR engagement and reporting. Marom (2017) introduces a conceptual framework to measure the CSR commitment of CFPs, which maps the stakeholders involved in a CF activity. The author labels as secondary stakeholders those that do not participate in the CF process directly, such as governments, employees and communities, while the primary stakeholders in crowdinvesting models are investors, shareholders, capital seekers and regulatory authorities. This study also itemises the CSR constituencies that a CFP, like any other business, should consider given its role of matching investors' financial resources with valuable projects.

The CSR rating of CFPs, asserts Marom, should be based on a set of issues related, for example, to transparent information for both investors and capital seekers, proper disclosure of information by CF proponents, and a corporate governance mechanism to protect CF stakeholders.

In line with the concept of the democratisation of finance, Ramos (2018) claims that, in connecting social capital with economic capital, CFPs are encouraging democratic and open innovation practices, with their campaign promotions based on values such as transparency, participation, learning communities and efficacy. The author noted that CFPs are characterised by a distinctive management style (which he names "efficiencracy"), which is based on the idea that both democratic practices and efficiency are necessary and complementary conditions for the success of campaigns and the prosperity of the CF market. Other arguments claiming a socially oriented role for CFPs are founded on positive externalities such as self-employment, new venture support, and the creation of new professionality and skills related to CFPs' business models (Ramos 2018; Mollick and Kuppuswamy 2014). Available studies on CSR–CFPs show that CFPs' engagement with CSR might be a factor affecting the long-term viability of these digital businesses. Indeed, a greater attention to CSR—and this is the case also for professional investors (Gangi et al. 2019)—can, on the side of the CFPs, promote a sustainable reputation and visibility, while also allocating financial resources to sustainability-oriented projects.

Moving from theory to practice, we can see that some CFPs have started to follow mainstream sustainable finance trends by considering CSR-related issues and the United Nations Development Programme's 2030 Agenda (the 17 Sustainable Development Goals)[2] as pillars of their efficiency over the long term. Produzioni dal Basso (PdB), an Italian reward–donation CFP that promotes social innovation, integrated into its campaign layout one or more of the SDGs in accordance with their relevance to project characteristics or expected outcomes (see Table 4.1). This informational signal raises crowd awareness about the social and environmental impacts of a CF project, favouring the allocation of resources

[2] "The Sustainable Development Goals (SDGs), also known as the Global Goals, were adopted by all United Nations Member States in 2015 as a universal call to action to end poverty, protect the planet and ensure that all people enjoy peace and prosperity by 2030." United Nations Development Programme (UNDP): https://www.undp.org/content/undp/en/home/sustainable-development-goals.html.

Table 4.1 CF and CSR: CF projects and the Sustainable Development Goals

SDGs	*Projects (%)*	*Funds raised (%)*	*Community (%)*
SDG1: No Poverty	5	6	4
SDG2: Zero Hunger	3	2	1
SDG3: Good Health and Well-being	11	10	7
SDG4: Quality Education	11	8	5
SDG5: Gender Equality	6	8	5
SDG6: Clean Water and Sanitation	1	1	0
SDG7: Affordable and Clean Energy	1	0	0
SDG8: Decent Work and Economic Growth	7	5	4
SDG9: Industry, Innovation and Infrastructure	2	1	1
SDG10: Reduced Inequalities	14	19	25
SDG11: Sustainable Cities and Communities	11	11	10
SDG12: Responsible Consumption and Production	6	5	4
SDG13: Climate Action	5	5	5
SDG14: Life below Water	1	0	0
SDG15: Life on Land	8	7	6
SDG16: Peace, Justice and Strong Institutions	5	8	19
SDG17: Partnerships for the Goals	3	3	2
Total	100	100	100

Source Authors' elaboration of publicly available Produzioni dal Basso information on CF donation- and reward-based campaigns. The web-scraping tool Octoparse was used to gather financial and ESG-oriented information on 2485 projects. The 2485 projects comprised a total volume of €8,360,707 involving over 106,000 crowdlenders

towards those projects that promote sustainable approaches or outcomes. An analysis of information on 2485 projects publicly hosted on the PdB website[3] reveals that projects featuring SDGs mainly (55% of projects) focus on those goals concerning the reduction of inequality (SDG 10),

[3] Authors' elaboration of publicly available PdB data. The PdB website has a section listing 2485 projects and providing information on how the project or its outcome is linked to the SDGs (https://www.produzionidalbasso.com). Data on projects were gathered with the web-scraping tool Octoparse and processed according to the following dimensions: SDGs, funds raised and community involved.

health and well-being (SDG 3), education (SDG 4), and sustainable cities and communities (SDG 11). In terms of funds raised, the majority of finance collected (55%) was for projects promoting reduction of inequality (SDG 10), and sustainable cities and communities (SDG 11). In terms of crowd participation, in addition to the SDGs already mentioned the pillar related to gender equality (SDG 5) registered highly.

Lendahand[4] is a pro-social lending CFP with a mission to attract finance entrepreneurs from underdeveloped countries, creating new jobs and access to basic needs such sanitary equipment and renewable energy with the wider aim of eradicating poverty. This CFP, unlike most other CFPs that promote pro-social lending with the related crowdlender motivations (e.g. Kiva; Berns et al. 2020; Allison et al. 2015), also provides a detailed disclosure of environmental and social performance in terms of the outcome of the loans financed. The information provided, as well as financial (loan amounts, maturity, annual interest rate, etc.), concerns the creation of new employment and sanitary installations (toilets), and, from an environmental perspective, where the loans are used to support system-related investments in renewable energy (see Table 4.2). Finally, it is worth noting that some CFPs are verticalised according to specific sustainability topics. Ecrowd! is a lending CF platform that supports socially innovative projects related to the circular economy, green energy, natural resources, health and nutrition, education and community. Lending CF projects posted on this platform must make clear in their initial proposals the intended positive impact to be generated in these areas. Trine connects crowd communities of investors with lending CF investment opportunities in solar energy in growing markets, tracking ESG progress to create a virtuous cycle of information about how the issuers' loans are contributing to communities' wealth and development.

Flórez-Parra et al. (2020), taking a sample of 101 SMEs seeking capital from private and institutional investors via a Spanish platform, investigate the power of the company's CSR rating to affect its CF success. In line with recent studies (Gangi et al. 2020a, b), that of Flórez-Parra et al. (2020) confirms that, in a CF mechanism, CSR engagement improves a CFP's ability to forge trusting relationships with financial investors that lower the cost of the finance, in turn positively impacting profitability. Therefore, those CFPs that use ESG ratings or a similar framework allow

[4] https://www.lendahand.com/en-EU.

Table 4.2 Social impact lending CF and sustainable performance ("spillovers")

Loans observed (1959) *Type of investment: direct (7%)* *Type of investment: local partner (93%)*	*Mean*	*St. dev.*	*Min*	*Max*
Loan amount	27,927.24	47,595.51	1,800.00	500,000.00
Maturity (months)	25.24	12.81	6	48
AIR (annual interest [%])	3.73	1.03	3	9
Crowdlenders	72.94	99.47	6	935
ENV SPILLOVER (solar energy system)	463.18	814.67	2	5000
ENV SPILLOVER (biodigester)	235.36	125.20	60	500
ENV SPILLOVER (green energy [KWh])	118.90	126.30	12	650
ENV SPILLOVER (tons of CO_2 reduced)	10,366.91	8,872.23	13	25,000
SOCIAL SPILLOVER (toilets)	117.12	169.07	23	910
SOCIAL SPILLOVER (jobs)	4.374776	5.151,941	1	80

Source Authors' elaboration of publicly available Lendahand.com information on pro-social crowdlending loans. The web-scraping tool Octoparse was used to gather financial and ESG-oriented information on 1959 loans. The 1959 loans comprised a total volume of €54,597,750, involving over 130,000 crowdlenders and 976,302 stakeholders such as financial institutions, local communities and entrepreneurs

crowdlenders to make a more accurate assessment of the impacts on the community and the environment generated by the investments. These signals, backed up by the financial advisor or a local financial partner (e.g. a bank, a microfinance institution) or mentor, create the conditions in which virtuous cycles of information and collaborative practices are activated. Covering a shortfall in financial resources through the integration of different assets, this results in both economic and social benefits for organisations, the financial institutions involved, individual investors and in fact the sum total of all actors involved (Flórez-Parra et al. 2020; Ramos 2018; Marom 2017).

4.4.1.2 Crowdfunding as a Stakeholder Engagement Model

In line with van Marrevijk's (2003) conceptualisation of CSR, and from a stakeholder theory perspective (Freeman 1984), CF can be investigated as a stakeholder management tool, supporting organisations in the digital era in responding more adeptly to stakeholders' CSR demands.

Andriof and Waddock (2002) define stakeholder engagement as a process for adequately managing a business's operational and reputational risks. Stakeholder management processes are tools with which management actively seeks dialogue with relevant parties to define a system of shared values regarding sustainability issues of common interest, while pursuing both economic and social goals. So, from the perspective of CSR engagement, organisations can employ CF to measure the degree of urgency assigned by a range of target stakeholders orbiting the CFP to business-related social and environmental issues and strategies.

In line with the resource-based view (Barney 1991) and knowledge management perspectives (Darroch 2005; Gangi et al. 2019), recent studies have demonstrated that a higher level of sustainability performance increases organisations' ability to exchange resources with stakeholders, such as knowledge and trust, leading to higher corporate reputation and financial performance (Gangi et al. 2020; Branco and Rodriguez 2006). In pursuit of reputational capital, organisations might have recourse to CF as a vehicle to communicate with the crowd about CSR initiatives, thus raising awareness of the company's CSR standing and increasing its reputation in the eyes of stakeholders (Spanos 2018). By participating in CF, organisations can send a commitment signal to the crowd about specific social or environmental issues, positioning them as co-financiers of and ambassadors for the value of a CF proposal. This increases the perception of an organisation as socially involved with respect to issues that the crowd consider worthy of support. Additionally, CF fosters transparency about the degree to which organisations actually care about and act in pursuit of sustainability-related issues (Spanos 2018).

It is useful to note that CSR issues assume different configurations, depending on the sector (Perez and del Bosque, 2013). Some sectors are under particular scrutiny on account of their environmental footprint (e.g. airlines, petrochemicals), others for their impacts on social and wealth inequality (e.g. fashion, banking, food). Among the benefits that may derive from media exposure, it is worth mentioning the possibility for organisations to partner with other actors from within their principal CSR sectoral domains. Here, organisations can use CF as a tool to allocate their CSR financial resources to CSR causes promoted by other market agents—causes that overlap with their own, thereby furthering their CSR reputation.

CF is generally considered a new financial model that can bridge the financial gap for those organisations and individuals often excluded from traditional channels. In the CSR literature, a firm's size is seen as a predictor of its CSR engagement and performance. Firms with bigger cash flows or with resource slack perform better in responding to stakeholders' CSR demands; also, large firms tend to be more exposed to public scrutiny, and thus more proactive about CSR (Donaldson 2001; Brammer and Millington 2006). The power of CF to level up these conditions, expanding the range of financial sources for CSR investment through fund-matching with the crowd, makes it particularly relevant for companies with financial constraints, such as SMEs, but which do not wish to be excluded from participation in sustainability initiatives (Cheng et al. 2014).

Finally, the synergies between CF and CSR are also relevant to socially responsible human resource management (Greenwood 2007; Barrena-Martinez et al. 2018). CF can be used as a mechanism for greater employee involvement in bottom-up CSR-oriented projects. Employees might be involved in the selection of CF initiatives, for example, in line with the company's CSR goals, or they could challenge themselves to handle the actual proposition and promotion of CSR projects. In 2012, IBM launched the first corporate CFP, aimed at promoting and funding projects launched by employees, an initiative that generated significant innovations in terms of new products, a better workplace environment, proactivity and increased productivity (Marom et al. 2015; Ramos 2018).

4.4.1.3 *Banking and CF: Framing Synergies*

The banking system was profoundly affected by the financial crisis due to its involvement in the diffusion of high-risk financial products and because its functioning model encouraged excessive speculative behaviour by financial operators (Gangi 2013). As a consequence, the sector faced a long period of recovery from declining revenues and the loss of consumer trust, something that banks are trying to restore by increasing their sustainability orientation, integrating CSR principles into their operational financial ratings model, and adopting a new mindset around environmental and societal issues with regard to both internal and external communities (Scholtens 2009; Gangi et al., 2019). Concurrently, ICT developments and related emerging fintech solutions contributed to a redesign of the financial markets, making a significant impact on traditional financial business models, consumption patterns and expectations. So the current competitive scenario for banks involves, from one side, emerging competition coming from the fintech startups trying to conquer the digital financial market niche. On the other side is the so-called GAFAM[5] nexus, which due to their technological influence and capacity to provide new integrated financial solutions, also represents a threat for the banking industry.

This turbulence, which was also linked to policymakers' regulations,[6] forced banking institutions to adapt rapidly to changes in the markets. The global banking system continues to downsize in structural terms,[7] in its efforts to balance analogue and digital components to compete effectively in the market (Barbagallo 2018). Recent publications point out that collaboration is the preferred engagement strategy of incumbents towards fintech startups, including those offering CF solutions (EY 2019; Bank of Italy, 2017; Starteed 2019; Beck et al. 2016; Hornuf et al. 2020). A majority of incumbents have recognised in fintech startups an opportunity to explore new tools, capabilities and approaches to improve customer experience and, ultimately, revenues[8] (Accenture 2019; EY 2019).

The bank–CF relationship represents an emerging and controversial topic within CF studies, inasmuch as there are arguments supporting both the competitive and strategic views of this relationship. Some argue that the benefits for banks in collaborating with fintech startups relate to the possibility of innovations in digital banking models (Accenture 2019; Capgemini 2017; Anagnostopoulos 2018; Cole et al. 2019). Others argue that fintech's alternative fundraising solutions for organisations and individuals might conceivably supplant incumbents. Whether this turns out to be the case or not, data on CF solutions and on financial customers'

[5] The acronym GAFAM refers to the five biggest tech companies dominating the Internet: Google, Apple, Facebook, Amazon and Microsoft. These US companies are constantly on society's and the regulators' radar on account of their huge influence on digital markets, due to their large technological influence, massive databases of user data and strong financialisation.

[6] In Europe, the EU's Payment Services Directive 2 (PSD2) provides consumers with more choices for payment operations among a range of financial providers as well as banks. It gives banks the opportunity to develop platforms that integrate alternative and traditional financial services.

[7] From 2008 to 2018, banks in Europe decreased by 29% (EBF, 2019) and by 22% in the Americas (FDIC Bank Annual Historical Bank Data, accessed July 2020).

[8] As revealed by Accenture (2019), banks involved in open innovation practices with fintech players can earn rewards of over 40% in terms of market value.

behaviour[9] suggest that banks cannot remain neutral about the growth of alternative financial models such as CF. Entrepreneurs who are recurrent users of CF often justify their choice by evoking factors such as more transparency, flexibility, rapidity and affordability compared to traditional channels (Blaseg and Koetter 2016; Wald et al. 2019). Consequently, both banks and CFPs are now required to satisfy new retail and corporate expectations of faster and more flexible financial digital solutions.

Although CF currently represents the majority of financial services available (Cole et al. 2019), few crowdinvesting studies have focused on the relationship between CF and banks. Blaseg and Koetter (2016) demonstrate that, in credit crunch conditions when banks suffer shocks, German young ventures are more willing to switch to equity CF. Moreover, the ventures that are most likely to apply for CF are characterised by lower creditworthiness, high risk and high level of intangible assets. Tang et al. (2018), observing the US lending CF market, reveal that bank borrowers use lending CF to complement their loans when those are relatively small, or, in other cases, CF is used an alternative source to banks entirely. Spanos (2018), analysing a CFP co-created by a Greek commercial bank and the government, observes how CF might be effectively incorporated into banks' CSR strategies. Cole et al. (2019) find negative, positive and inconsistent relationships between bank finance and CF sources. The authors analysed banks' failure data and reward and lending CF campaigns from four leading CFPs. The results imply a complementary relationship between bank finance and crowdfunding: that is, bank defaults also cause a decrease in CF success rates in both the reward and lending models.

Essentially, then, the few existing studies on the nature of the relationship between banks and financial CF sources offer contrasting evidence. Also, the evolving relationship between banks and CFPs, and the difficulty in gathering reliable data (Blaseg and Koetter 2016; Spanos 2018), leave the areas of collaboration between banks and CF actors still to be explored, especially from the perspective of a CSR–CF banking relationship. Indeed, with the literature demonstrating the growing relevance of socially responsible banking to the stability of financial markets (Scholtens

[9] According to CCAF (2020), out of businesses that raised capital from equity CFPs, 25% did so more than once, while in the P2P business lending segment this proportion rises to 34%. The EY FinTech Adoption Index 2019 revealed that the global consumer fintech adoption rate is 64%, with this percentage dropping to 35% for SMEs.

2009; Gangi et al., 2019), a gap emerges on how banks are actually integrating CF as a model of stakeholder engagement and sustainable innovation.

With the aim of investigating how banks are interacting with CF actors and to what extent, below we explore, albeit tentatively, the usefulness and applicability of CF financial services in banks' business models, by drawing on 24 bank–CFP partnerships. We traced the initiatives of 17 European commercial banks and 20 CFPs in Europe and one collaboration from the US. In terms of corporate strategy, the majority of collaborations are related to business diversification (67%), 29% to CSR strategy and a small proportion (4%) to open innovation initiatives (Table 4.3).

Table 4.3 Banks and crowdfunding

Initiatives	*24*
Banks	23
CFPs	21
Initiatives by CF model (%)	
Equity	25%
Reward/Donation/Civic	54%
Lending and real estate	21%
Total	100%
Initiatives by strategy (%)	
Business-oriented	67%
CSR-oriented	29%
Open innovation	4%
Total	100%
Match-funding projects[a] *(% of total initiatives)*	
Number of projects	267
Success rate[a]	94%
Funds collected (€ million)	34
Average contributors[a]	672

[a]Match-funding project breakdowns represent data available from CFPs on 100 projects out of a total of 267, along with data on volumes and contributors as made available by CFPs or bank publications

Source Data were manually collected from a variety of public sources. Information on the 24 collaborations was collected from media and websites of the CFPs and banks. Where possible, data on projects supported by banks using CF were collected from the CFPs using the web-scraping tool Octoparse. In some cases, data were collected from CFPs' statistics and websites. Last update: February 2020

First, the bank–CF relationship can be understood in terms of business diversification opportunities for banks. In some cases, banks collaborate with CFPs with the aim of expanding their fintech product domain by integrating CF solutions into their products. In this case, banks serve as a connector with alternative funding possibilities ("indirect fund-matching"), subsequently expanding the funding options available to their clients via CFPs. In an indirect fund-matching process, banks use CF to offer their customers more flexible ways of obtaining resources in addition to, or as an alternative to, their own financial services. In Belgium, ING partnered with KissKissBankBank (a reward-based CF) with the aim of expanding the financing options available to its clients and their projects. In Spain, Evo Banco invites its customers to invest in property projects by means of a collaboration with the real estate CFP Housers. Further areas of complementarity between CF and banks' digital innovation ambitions could include the credit control process (Cole et al. 2019). A successful CF campaign is likely to serve as a credible signal about an entrepreneur when he/she applies for further rounds of capital (Hornuf et al. 2018). In a similar way, banks might consider prior CF experience as a factor in mitigating a firm's risk profile. Certainly, in terms of bankability, a successful prior CF campaign represents an additional guarantee of a project's viability to the extent that it demonstrates the existence of a target audience, its capacity to engage customers, and implies the acquisition of trust and reputation (Mollikc, 2014; Frydrych et al. 2016).

Further synergies that banks might look for in their collaborations with CFPs relate to the possibility of intercepting new investment opportunities and expanding clients' target markets by co-financing highly specific CF campaigns with direct relevance to the core business ("direct fund-matching"). In the direct fund-matching model, banks act as professional crowdfunders, or major contributors, supporting CF entrepreneurs and new ventures of interest to the banks on account of their core businesses.

The direct fund-matching method is significantly more likely to be used if the co-financing activity relates to projects and initiatives related to CSR strategy. In fact, 54% of direct fund-matching operations relate to CSR projects and open calls promoted or supported by banks, alongside the CFPs' virtual communities.

Finally, a third means of integration between CF domains and banks—analogous to how it works with professional investors—relates to the way CFPs serve as fertile ecosystems for the detection and screening of emerging technologies and partners for innovation in the fintech market.

In 2016, BNP Paribas and SmartAngels (an equity CFP) collaborated to create a digital report based on blockchain technology, in order to secure equities offered by equity CF. This type of collaboration aligns with the perspective of a complementary relationship between banks and CF (Cole et al. 2019), where banks gain competitive advantage by offering CFPs their long-term expertise in business due diligence, corporate financing and regulatory frameworks.

Aware of the intensity with which these forms of collaboration can occur, in some cases banks prefer to adopt in-house solutions by integrating tailored solutions from CF technology providers and thereby saving costs of internal development. ForFunding[10] is the donation-lending CF platform of Intesa Sanpaolo, which brings together individuals and businesses to collaborate financially via donations or community lending to projects put forward by organisations that have research, social or cultural purpose. As discussed at length in Chapter 3, the concept of socially responsible banking emerged in the field of CSR studies as a topic that aligned with a Schumpeterian perspective on bank responsibilities (Schumpeter, 1939, 1964). Due to their role as intermediaries between those who need capital and those wishing to invest, banks are crucial levers for economic development. Banks have a social role because they do not act on behalf of a restricted, predefined group of agents but on the behalf of society as a whole, being called on to allocate resources among the best entrepreneurial opportunities to promote collective wealth and development.

Relevant to this because of the bank's ethical business model orientation is the collaboration between Starsup, an Italian equity CFP, and Banca Etica, which acted as an institutional co-financier for social enterprises seeking capital via equity CFPs. In cases such as this, banks are experimenting with CF as a screening tool to expand their chances of tapping into investment opportunities that are in line with their business model. This generates opportunities for banks to come into contact with entrepreneurial proposals or startups that are valuable to them from the point of view of governance mechanisms and environmental and social commitments.

[10] https://www.forfunding.intesasanpaolo.com/DonationPlatform-ISP.

Indeed, banks' willingness to engage in CF is related to the chances they perceive they would have of intercepting new fund-matching opportunities that can make their support for sustainable development more tangible. In Italy, Banca Etica, in collaboration with the reward-based CFP Produzioni dal Basso, is co-financing CSR-oriented projects through its Etica SGR fund. In 2014, Société Générale, in partnership with SPEAR, financed €750,000 worth of projects with a high social and environmental value. BPER, in collaboration with PdB, is co-financing projects with a social mission. Santander, via the ChangeMaker fund, allocated over $1 million for social enterprise projects and nonprofit organisations presented on the crowdfunder reward-based CFP.

In line with the evolving scenario, banks are experimenting with CF as a digital CSR tool to establish direct relationships with local communities, consumers, workers and customers, favouring the alignment of interests between economic and social actors. The salience of the collective social capital dimension in the functioning of CF represents, for banks, a lever to make their commitment to stakeholders about social and environmental issues more tangible. By working alongside the crowd as co-financiers, banks offer third-party endorsement of proposed initiatives, while providing concrete financing solutions for those on the fringes and excluded from banking channels (Cumming et al. 2019; Spanos 2018). Their role as big donors and mentors means that they can intercept the issues that the crowd consider worthy of support and invest in them, thereby gaining reputation and credibility. Moreover, as highlighted by Spanos (2018), banks might expand their CSR financial options through CF via match-funding schemes involving the banks, other institutions and the crowd.

In conclusion, in observing these forms of alliances, we can see that, for banks, the rise of CF is a sign that customers expect simpler, more transparent and more flexible financing models that take digital technologies into account. By joining forces with CFPs, banks are looking for ways to meet these requirements. Indeed, with CF being a more participative model with which to accrue both financial and social capital, it represents for the banking industry a powerful tool to explore new areas of value creation in accordance with banks' economic and social responsibilities.

4.4.2 CF as a Form of Impact Finance

According to the United Nations,[11] there is an annual funding gap of around $3 trillion for its 2030 Agenda Sustainable Development Goals. Impact finance instruments are widely considered to have a key role to play in bridging this gap and, in a broader view, in achieving sustainability. According to Höchstädter and Scheck (2015), the term "impact finance" is defined with reference to two main concepts: the provision of a deliberate financial return for investors and a social/sustainability approach. With regard to the financial dimension, the impact investing philosophy is distinct from grant and venture philanthropy because of its expectations of financial return, even if lower than or in line with market rates. With regard to the social or sustainable dimensions of these investments, there is a conceptual overlap with business ethics, CSR and socially responsible investments.

Impact investments, therefore, aim to increase the quality and quantity of sustainable outcomes obtainable by business activities, with a specific focus on those activities that are expected to generate primary collective goods and benefits, such as creating good jobs, more ethical consumption models, environmentally friendly practices or responsible stakeholder governance (Choi et al. 2013; Höchstädter and Scheck 2015). To broaden the sources of financing for underbanked or unbanked individuals and organisations, impact investing is an appropriate model to consider, alongside other useful routes such as microfinance, venture philanthropy (Frumkin 2003; Perrini and Vurro 2010), public–private social collaborations and CF (Lehner 2013; Lehner and Nicholls 2014; Balboni et al. 2016; Cumming et al. 2019), the latter allowing many-to-many relationship among organisations and different types of financial stakeholders. The CCFA report (2020) indicates the relevance of crowd-investing models for non-bankable individuals, especially in the P2P lending market, where the majority of applicants appear to be underbanked or unbanked.

From a financial inclusion perspective, CF is seen as a model to support businesses and entrepreneurs that are excluded from traditional financial channels for any of a number of reasons, such as geographical location, ethnicity, gender or low-wage conditions (McGuire, 2017; Marom et al. 2015; Vismara 2016; Cumming et al. 2019). From a gender perspective,

[11] https://www.un.org/sustainabledevelopment/sg-finance-strategy.

McGuire (2017) empirically demonstrates that equity CF has a positive impact on female startups' access to external capital, and Vismara (2016) find that successful campaigns are more likely to have a female CEO heading the firm, and that, on average, women tend to invest more than men, even though they are usually outnumbered. In the reward-based CF model, however, Marom et al. (2015) find that, even though female CF proponents have a higher success rate, the gender bias is noticeable in terms of funds raised, which are higher for male entrepreneurs. Nonetheless, reward-based CF is disproportionately popular with female investors, who mostly tend to finance projects led by females, which also concurs with the findings of Vismara (2016) in relation to equity CF. Bento et al. (2019) also find, in reward-based CF, that higher female ratios in the venture teams is correlated with better success rates after the CF campaign. Cumming et al. (2019) find equity CF a powerful instrument allowing financial minorities to interact with different types of financial interlocutor, but they find no evidence about the role of equity in bridging the gender gap in terms of fundraising. The capacity of CF to democratise access to financial resources is clearer with regard to remotely located companies, which benefit from CF's potential to communicate proposals worldwide.

Social enterprises—and socially oriented entrepreneurs—represent another category traditionally insufficiently served by traditional channels, despite their social potential; for these, impact investing is often seen as a panacea to overcome financial barriers. Although the definition of a social enterprise can change according to the regulatory and cultural context (Doherty et al. 2014), the literature describes them as hybrid organisations that use social, governance and environmental sustainability concepts to create new solutions in the private market (Doherty et al. 2014; Dart 2004; Mair and Martí 2006). In line with the European Commission (2015, p. 9), what distinguishes social enterprises from mainstream enterprises is their pursuit of a commercial activity directly linked to the provision of societal objectives or common goods, the profit from which is reinvested in the achievement of its social mission; they are also distinguished by participatory and fair governance models with regard to customers, employees, suppliers, stakeholders and government. From a financial standpoint, these organisations tend to combine a plurality of sources, both primary and subsidiary in nature, such as impact bonds, public funds and grants and partnerships with ethically oriented banks and mainstream financial institutions. However, a number of studies have highlighted how traditional forms of financial support for economic development are not able to meet social enterprises' needs (Ridley-Duff, 2008; Fedele and Miniaci 2010). The problem is mainly linked to cognitive and cultural barriers in the dialogue between social enterprises and traditional operators, which limit the ability of the latter to fully evaluate these businesses' risk and potential, resulting in a failure to properly integrate social and environmental impact into investment assessments (Lehner 2013; Brown 2006; Ridley-Duff, 2008).

Lehner and Nicholls (2014) advance CF's potential to enlarge the audience for social investment, by observing public–private partnership models between government, business and socially oriented investors, which are enhancing social enterprises' ability to raise risk capital through equity CF. As well as leveraging CF's power to combine the values of trust, transparency and collaboration (Lehner 2013) with financial expectations, social enterprises can find among the crowds those investors less inclined to invest in conventional, high-risk activities such as polluting industries (Lehner and Nicholls 2014). Further evidence is provided by Balboni et al. (2015, 2016), who, in analysing the success factors of donation- and reward-based CF, also note CF's capacity to increase the legitimacy of social enterprises' innovation projects.

CF, as a model for collective participation, innovates both financial and stakeholder engagement for entrepreneurs. This applies not only to traditional entrepreneurship but also—and perhaps even more so—for unrepresented entrepreneurs and social enterprises. In summary, CF has potential not simply because it is a financing technique based on digital technologies, but because it can be a tool to build social capital, fuelling social responsibility at both organisational and community level. What makes this alternative financial model consistent with certain concepts of social responsibility and sustainability is the set of intangible resources that make it an elective financial medium for the generation of wider value through financial support and inclusion.

4.4.3 *Sustainability-Oriented CF Projects and Crowd Behaviour*

As discussed, the financing of innovation and entrepreneurial initiatives has been shaped by the emergence of CFPs, through which individuals and organisations collaborate to finance projects (Cumming and Hornuf 2018). On account of the way it functions—an open network disseminating social innovation—CF can break down cultural barriers in the financing of sustainability. The rapid growth of CF is helping to establish a new crowd economy model based on sustainable values. Previously, only those with substantial cultural and financial capital were likely to engage with sustainable innovation projects. With CF, the visibility of such valuable proposals can potentially reach a global audience, empowering the "wisdom of the crowd" and participation in socially oriented investments. Studies on sustainable CF (Bento et al. 2019; Flórez-Parra et al. 2020) show us that this phenomenon is changing the financial markets by bringing together organisations and socially responsible investors from the crowd, and driving social change via economic, social and environmental opportunities.

Some authors have observed how sustainability-related signals are influencing crowd behaviour with regard to their financial support and motivations within different CF models. Mastrangelo et al. (2019) investigated how the CSR commitments of proponents relate to their success in raising capital. They empirically demonstrate CF's empowerment of sustainability-oriented projects, showing that the crowd reward entrepreneurs with a concern for CSR issues related to governance, employees and consumers. Calic and Mosakowski (2016), focusing on reward-based CF, find that the sustainable orientation of a project is equated with higher funding success, and that this effect is stronger for highly creative or third-party-endorsed projects. Similarly, Wehnert et al. (2019) analysed the sustainability signals associated with products offered via CF under the assumption that sustainability information drives a CF campaign's success. Their results show that, generally speaking, the credibility of sustainability features such as environmental friendliness and social fairness will increase success; however, this effect diminishes as the complexity of the product increases. In contrast, Testa et al. (2020), who focused on the sustainable attributes of food products pre-purchased via Kickstarter, assert that, while innate product characteristics can drive a CF campaign's success, an emphasis on society-centred motives is of greater use for CF campaigns with local implications and local benefits.

In the donation-based CF segment, Behl and Dutta (2019) focused on the links between CSR and CF in the context of humanitarian operations, investigating the relationship between the level of information disclosed by proponents and financial and social support. They find that CSR activities enhance financial support for social causes and also create a niche for dialogue and further collaboration with different stakeholders.

Berns et al. (2020) define pro-social lending CF as a form of CF in which crowdlenders can invest, providing debt without interest; they go on to investigate whether such lenders are driven by altruistic or strategic reasons. Through the lens of ethical and socially oriented lending, the authors took a large sample from Kiva CFPs and demonstrated that success in the funding of pro-social loans is driven primarily by a strategic rather than altruistic motivation. This aligns with Pierrakis's study (2019), which shows that lenders tend to be more expert in a lending CF context than in other models. However, Allison et al. (2015), also looking at altruistic narratives in pro-social lending, show that fully funded projects are characterised by higher levels of both social and strategic information. The inference is that the crowd tends to be rational in avoiding higher-risk borrowers and chooses high-quality entrepreneurs in order to retain the opportunity to reinvest their capital in other valuable entrepreneurs in the future. Berns et al. (2020), nonetheless, point out that altruistic narratives as well as the appeal of the entrepreneurial profile both have a bearing on the amounts loaned by crowdlenders.

Vismara (2019) analysed the impact of the sustainability orientation of equity CF offerings and established that, even though sustainability is not a driver of success in and of itself, CSR engagement and communication from equity-backed ventures are likely to attract and engage retailer investors—namely, those generally investing relatively small amounts—and that these investors, unlike professional investors, are more likely to bring a communitarian logic to their investment rather than applying a market perspective.

Finally, CF is emerging as an ideal source of financing for green projects, i.e. those with environmentally friendly and innovative solutions. According to prior studies (Kunze and Becker 2014; Butticè et al. 2019), CF is a powerful communication and aggregation tool that can favour participation and democratisation in addressing the energy sector's future challenges, especially in those countries with limited environmental engagement. The green projects more likely to find funding are those that bring economic benefit to the local community. The effects of green CF are mainly manifested in improved quality of life at a local level due to improved environmental performance, such as air quality, biodiversity and habitat, water and sanitation. Green CF campaigns also have a positive effect on general popular awareness about the environment and sustainability. Bonzanini et al. (2015) claim that CF is a valuable source of capital for green initiatives, as it combines the opportunity to generate profit with a moral purpose and the imperative to take action on climate change. However, looking at it from an agency theory perspective, Hörisch (2020), in investigating the performance of environmentally oriented CF companies, reveals that only a few CF creators disclose a sufficient amount of information about how CF capital is allocated towards the fulfilment of the declared goals, leaving many crowdfunders unable to fully evaluate how environmental claims are being addressed.

Generally speaking, the CF studies cited report evidence in alignment with studies on the motivations and incentives of socially responsible and ethical investors (Berns et al. 2020). However, the contrasting evidence about whether a socially responsible orientation is a driver of CF success implies that, even though CF can support the financing of sustainable projects, effective assessment of sustainability issues and future outcomes is constrained, as it is in mainstream finance, by strategic and information barriers. Nonetheless, CSR engagement and communication is looking increasingly like a driver of success in the CF context, promoting sustainable financing and innovative market solutions (Bento et al. 2019; Flórez-Parra et al. 2020).

Finance has a critical role to play in building a more sustainable, healthy and resilient future, and CF can be a valuable tool to involve the general public in activating socially oriented market proposals. Impact investing instruments, such as CF, can offer a more effective response than traditional models to the growing demand for financial inclusion from social enterprises, socially oriented projects and non-bankable individuals. While ethically or ESG-oriented financial instruments are suited to professional investors and funds, crowd-based models represent a powerful tool to increase public engagement and further positive economic, social and environmental sustainability objectives.

References

Accenture (2019) Banche, parola d'ordine fintech: in Europa 400 accordi con le startup, Wired, Accessed 27 May 2019

Agrawal A, Catalini C, Goldfarb A (2011) The geography of crowdfunding (NBER Working paper No. 16820). https://doi.org/10.3386/w16820

Agrawal A, Catalini C, Goldfarb A (2014). Some simple economics of crowdfunding. Innov Policy Econ 14(1):63–97. https://doi.org/10.1086/674021

Agrawal A, Catalini C, Goldfarb A (2015) Crowdfunding: geography, social networks, and the timing of investment decisions. J. Econ. Manag. Strategia 24(2):253–274. https://doi.org/10.1111/jems.12093

Ahlers GK, Cumming D, Günther C, Schweizer D (2015). Signaling in equity crowdfunding. Entrep Theory Pract 39(4):955–980. https://doi.org/10.1111/etap.12157

Akerlof GA (1970). The market for "lemons": quality uncertainty and the market mechanism. Q J Econ 84(3):488–500. https://doi.org/10.1016/b978-0-12-214850-7.50022-x

Allison TH, Davis BC, Short JC, Webb JW (2015) Crowdfunding in a prosocial microlending environment: Examining the role of intrinsic versus extrinsic cues. Entrep Theory Pract 39(1):53–73. https://doi.org/10.1111/etap.12108

Anagnostopoulos I (2018) Fintech and regtech: Impact on regulators and banks. Journal of Economics and Business 100:7–25. https://doi.org/10.1016/j.jeconbus.2018.07.003

Andriof J, Waddock S (2002, December 26) Unfolding stakeholder engagement. Unfolding stakeholder thinking: Theory Responsib Engagem 1(121):17–42

Angerer M, Brem A, Kraus S, Peter A (2017) Start-up funding via equity crowdfunding in Germany: A qualitative analysis of success factors. J Entrep Finance 1:1–34

Ariely D, Simonson I. (2003) Buying, Bidding, Playing, or Competing? Value Assessment and Decision Dynamics in Online Auctions. J Consum Psychol 13(1&2):113–123. https://doi.org/10.1207/s15327663jcp13-1&2_10

Axjonow A, Ernstberger J, Pott C (2018) The impact of corporate social responsibility disclosure on corporate reputation: a non-professional stakeholder perspective. J Bus Ethics 151(2):429–450. https://doi.org/10.1007/s10551-016-3225-4

Azar S, Mackey TK (2015) Crowdfunding: A New Untapped Opportunity for Biotechnology Start-ups? J Commer Biotechnol 21(5):15–28. https://doi.org/10.5912/jcb717

Balboni B, Kocollari U, Pais I (2015) I segreti del successo delle campagne di crowdfunding delle imprese sociali italiane. Impresa Sociale 6:6–16

Balboni B, Kocollari U, Pais I (2016) Crowdfunding for Social Enterprises: An Exploratory Analysis of the Italian Context. International Perspectives on Crowdfunding, Emerald Group Publishing Limited, 65–79. https://doi.org/10.1108/978-1-78560-315-020151004

Banca d'Italia (2017) FINTECH IN ITALIA. Indagine conoscitiva sull'adozione delle innovazioni tecnologiche applicate ai servizi finanziari

Barbagallo C (2018) Il sistema bancario italiano: situazione e prospettive [Italy's banking system: The current situation and the outlook]. Speech by the

Director General for Financial Supervision and Regulation at the ASSBB, Varignana (Bologna–Italy)

Barnett ML (2007) Stakeholder influence capacity and the variability of financial returns to corporate social responsibility. Acad Manage Rev 32(3):794–816. https://doi.org/10.5465/amr.2007.25275520

Barnett ML, Salomon RM (2012) Does it pay to be really good? Addressing the shape of the relationship between social and financial performance. Strateg Manag J 33(11):1304–1320. https://doi.org/10.1002/smj.1980

Barney J (1991) Firm resources and sustained competitive advantage. J Manage 17(1):99–120. https://doi.org/10.1177/014920639101700108

Barrena-Martinez J, López-Fernández M, Romero-Fernandez PM (2018) Drivers and barriers in socially responsible human resource management. Sustainability 10(5):1532. https://doi.org/10.3390/su10051532

Basel Committee on Banking Supervision (2018) Sound Practices - Implications of fintech developments for banks and bank supervisors

Battisti E, Creta F, Miglietta N (2020) Equity crowdfunding and regulation: implications for the real estate sector in Italy. J Financial Regul Compliance, https://doi.org/10.1108/jfrc-08-2018-0109

Beaulieu T, Sarker S, Sarker S (2015) A conceptual framework for understanding crowdfunding. Communications of the Association for Information Systems 37(1):1. https://doi.org/10.17705/1cais.03701

Beck R, Brüntje A, Dardour O, Gajda D, Marom IP, Will A (2016) Introduction. Pages 1–4 in D. Brüntje and O. Gajda, editors. Crowdfunding in Europe, FGF studies in small business and entrepreneurship. Springer International, Switzerland

Behl A, Dutta P (2019) Social and financial aid for disaster relief operations using CSR and crowdfunding. Benchmarking 27(2):732–759. https://doi.org/10.1108/bij-08-2019-0372

Belleflamme P, Lambert T, Schwienbacher A (2014) Crowdfunding: Tapping the right crowd. J Bus Ethics 29(5):585–609. https://doi.org/10.1016/j.jbusvent.2013.07.003

Bento N, Gianfrate G, Thoni MH (2019) Crowdfunding for sustainability ventures. J Clean Prod 237:117751. https://doi.org/10.1016/j.jclepro.2019.117751

Berns JP, Figueroa-Armijos M, da Motta Veiga SP, Dunne TC (2020) Dynamics of lending-based prosocial crowdfunding: Using a social responsibility lens. J Bus Ethics 161(1):169–185. https://doi.org/10.1007/s10551-018-3932-0

Bhattacharya CB, Sen S (2004) Doing better at doing good: when, why, and how consumers respond to corporate social initiatives. Calif Manag Rev 47(1):9–24. https://doi.org/10.2307/41166284

Bi S, Liu Z, Usman K (2017) The influence of online information on investing decisions of reward-based crowdfunding. J Bus Res 71:10–18. https://doi.org/10.1016/j.jbusres.2016.10.001

Blaseg D, Koetter M (2016) Crowdfunding and bank stress. In Banking beyond banks and money, 17–54. Springer, Cham. https://doi.org/10.1007/978-3-319-42448-4_3

Block J, Hornuf L, Moritz A (2018) Which updates during an equity crowdfunding campaign increase crowd participation? Small Bus Econ 50:3–27. https://doi.org/10.1007/s11187-017-9876-4

Bonzanini D, Giudici G, Patrucco A (2016) The crowdfunding of renewable energy projects. In Handbook of environmental and sustainable finance, 429–444. Academic Press. https://doi.org/10.1016/b978-0-12-803615-0.00021-2

Bottiglia R, Pichler F (2016) Crowdfunding for SMEs. Springer, Macmillan Publishers Ltd. London. https://doi.org/10.1057/978-1-137-56021-6_2

Brabham DC (2008) Moving the Crowd at iStockphoto: The Composition of the Crowd and Motivations for Participation in a Crowdsourcing Application. First Monday 13(6):1–22. https://doi.org/10.5210/fm.v13i6.2159

Brammer S, Millington A (2006) Firm size, organizational visibility and corporate philanthropy: An empirical analysis. Bus Ethics 15(1):6–18

Brammer SJ, Pavelin S (2006) Corporate reputation and social performance: The importance of fit. J Manage studies 43(3):435–455. https://doi.org/10.1111/j.1467-6486.2006.00597.x

Branco MC, Rodrigues LL (2006) Corporate social responsibility and resource-based perspectives. J Bus Ethics 69(2):111–132. https://doi.org/10.1007/s10551-006-9071-z

Bretschneidera U, Leimeisterab JM (2017) Not just an ego-trip: Exploring backers' motivation for funding in incentive-based crowdfunding. J Strateg Inf Syst 26(4):246–260. https://doi.org/10.1016/j.jsis.2017.02.002

Brown J (2006) Equity finance for social enterprises. Soc Enterp J 2(1):73–81. https://doi.org/10.1108/17508610680000714

Bruton G, Khavul S, Siegel D, Wright M (2014). New Financial Alternatives in Seeding Entrepreneurship: Microfinance, Crowdfunding, and Peer–to–Peer Innovations. Entrep Theory Pract 39(1):9–26. https://doi.org/10.1111/etap.12143

Butticè V, Colombo MG, Wright M (2017) Serial Crowdfunding, Social Capital, and Project Success. Entrep Theory Pract 41(2):183–207. https://doi.org/10.1111/etap.12271

Butticè, Vincenzo, Massimo G. Colombo, Elena Fumagalli, and Carlotta Orsenigo. "Green oriented crowdfunding campaigns: Their characteristics and diffusion in different institutional settings." Technological Forecasting and Social Change 141 (2019): 85–97

Calic G, Mosakowski E (2016) Kicking off social entrepreneurship: How a sustainability orientation influences crowdfunding success. J Manage Studies 53(5):738–767. https://doi.org/10.1111/joms.12201

Capgemini BNP (2017). Paribas (2017). World payments report

CCAF - Cambridge Centre for Alternative Finance (2020) Global Alternative Finance Market Benchmarking Report - Trends, Opportunities and Challenges for Lending, Equity, and Non-Investment Alternative Finance Models

Chandler JD, Vargo SL (2011) Contextualization and value-in-context: How context frames exchange. Mark Theory 11(1):35–49. https://doi.org/10.1177/1470593110393713

Chen Y, Xie J (2005) Third-Party Product Review and Firm Marketing Strategy. Mark Sci 24 (2):218–40. https://doi.org/10.1287/mksc.1040.0089

Cheng B, Ioannou I, Serafeim G (2014) Corporate social responsibility and access to finance. Strateg Manag J 35(1):1–23. https://doi.org/10.1002/smj.2131

Choi BB, Lee D, Park Y (2013) Corporate Social Responsibility, Corporate Governance and Earnings Quality: Evidence from Korea. Corp Gov 21(5):447–467. https://doi.org/10.1111/corg.12033

Cholakova M, Clarysse B (2015) Does the Possibility to Make Equity Investments in Crowdfunding Projects Crowd Out Reward–Based Investments? Entrep Theory Pract 39(1):145–172. https://doi.org/10.1111/etap.12139

Chu CC, Cheng YF, Tsai FS, Tsai SB, Lu KH (2019) Open Innovation in Crowdfunding Context: Diversity, Knowledge, and Networks. Sustainability 11:180. https://doi.org/10.3390/su11010180

Cole RA, Cumming DJ, Taylor J (2019) Does FinTech Compete With or Complement Bank Finance? https://doi.org/10.2139/ssrn.3302975

Colombo MG, Franzoni C, Rossi–Lamastra C (2015) Internal Social Capital and the Attraction of Early Contributions in Crowdfunding. Entrep Theory Pract 39(1):75–100. https://doi.org/10.1111/etap.12118

Connelly BL, Certo ST, Ireland RD, Reutzel CR (2011) Signaling Theory: A Review and Assessment. J Manage 37(1):39–67. https://doi.org/10.1177/0149206310388419

Cordova A, Dolci J, Gianfrate G (2015) The Determinants of Crowdfunding Success: Evidence from Technology Projects. Procedia Soc Behav Sci 181(11):115–124. https://doi.org/10.1016/j.sbspro.2015.04.872

Courtney C, Dutta S, Li Y (2017) Resolving Information Asymmetry: Signaling, Endorsement, and Crowdfunding Success. Entrep Theory Pract 41(2):265–290. https://doi.org/10.1111/etap.12267

Cova B, Dalli D, Zwick D (2011) Critical perspectives on consumers' role as 'producers': Broadening the debate on value co-creation in marketing processes. Mark Theory 11(3):231–241. https://doi.org/10.1177/1470593111408171

Cumming D, Hornuf L (2018) The economics of crowdfunding. Palgrave Macmillan

Cumming D, Johan S, Zhang Y (2018) Public policy towards entrepreneurial finance: spillovers and the scale-up gap. Oxford Review of Economic Policy 34(4):652–675. https://doi.org/10.1093/oxrep/gry012

Cumming D, Meoli M, Vismara S (2019) Investors' choices between cash and voting rights: Evidence from dual-class equity crowdfunding. Res Policy 48(8):103740. https://doi.org/10.1016/j.respol.2019.01.014

Dahlsrud A (2008) How corporate social responsibility is defined: An analysis of 37 definitions. Corp Soc Responsib Environ Manag 15(1):1–13. https://doi.org/10.1002/csr.132

Darroch J (2005) Knowledge management, innovation and firm performance. J Knowl Manag 9(3):101–115. https://doi.org/10.1108/13673270510602809

Dart R (2004) The Legitimacy of Social Enterprise. Non-profit Management and Leadership 14(4):411–424. https://doi.org/10.1002/nml.43

Davis MT (2015) Crowd Actions in Britain and France from the Middle Ages to the Modern World. Palgrave Macmillan

De Bakker FG, Hellsten I (2013) Capturing online presence: Hyperlinks and semantic networks in activist group websites on corporate social responsibility. J Bus Ethics 118(4):807–823. https://doi.org/10.1007/s10551-013-1962-1

De Buysere K, Gajda O, Kleverlaan R, Marom D (2012) A framework for European crowdfunding. Available at: http://www.crowdfundingframework.eu/

De Luca VV, Margherita A, Passiante G (2019) Crowdfunding: a systemic framework of benefits. International Int J Entrepreneurial Behav Res 25(6):1321–1339. https://doi.org/10.1108/ijebr-11-2018-0755

Di Domenico M, Haugh H, Tracey P (2010) Social Bricolage: Theorizing Social Value Creation in Social Enterprises. Entrep Theory Pract 34 (4):681–703. https://doi.org/10.1111/j.1540-6520.2010.00370.x

Doherty B, Haugh H, Lyon F (2014) Social enterprises as hybrid organizations: A review and research agenda. Int J Manag Rev 16(4):417–436. https://doi.org/10.1111/ijmr.12028

Donaldson L (2001) The contingency theory of organizations. Sage

Drover W, Busenitz L, Matusik S, Townsend D, Anglin A, Dushnitsky G (2017). A Review and Road Map of Entrepreneurial Equity Financing Research: Venture Capital, Corporate Venture Capital, Angel Investment, Crowdfunding, and Accelerators. J Manage 43(6):1820–1853. https://doi.org/10.1177/0149206317690584

Esposito RT (2012) The social enterprise revolution in corporate law: A primer on emerging corporate entities in Europe and the United States and the case for the benefit corporation. Wm & Mary Bus L Rev 4:639. Doi:

Estellés-Arolas E, González-Ladrón-de-Guevara F (2012) Towards an integrated crowdsourcing definition. J Inf Sci 38(2):189–200. https://doi.org/10.1177/0165551512437638

European Commission (2015) A map of social enterprises and their eco-systems in Europe Synthesis Report

EY (2019) Real Estate Crowdfunding - Introduction to an alternative way of investing

Fedele A, Miniaci R (2010) Do social enterprises finance their investments differently from for-profit firms? The case of social residential services in Italy. J Soc Entrepreneurship 1(2):174–189. https://doi.org/10.1080/19420676.2010.511812

Fisher E (2011) Editorial overview. Science and engineering ethics 17(4):607–620. https://doi.org/10.1007/s11948-011-9331-x

Flórez-Parra JM, Rubio Martín G, Rapallo Serrano C (2020) Corporate Social Responsibility and Crowdfunding: The Experience of the Colectual Platform in Empowering Economic and Sustainable Projects. Sustainability 12(13):5251. https://doi.org/10.3390/su12135251

Fombrun CJ, Gardberg NA, Sever JM (2000) The Reputation QuotientSM: a multi-stakeholder measure of corporate reputation. J Brand Manag 7(4):241–255. https://doi.org/10.1057/bm.2000.10

Foster RJ (2011) The Uses of Use Value: Marketing, Value Creation, and the Exigencies of Consumption Work', in D. Zwick and J. Cayla (eds) Inside Marketing: Practices, Ideologies, Devices, pp. 42–57. Oxford and New York: Oxford University Press

Freeman RE (1984) Strategic Management: a Stakeholder Approach. Pitman, Boston

Frumkin H (2003) Healthy places: exploring the evidence. Am J Public Health, 93(9):1451–1456

Frydrych D, Bock AJ (2018). Bring the Noize: Syndicate and Role-Identity Co-Creation During Crowdfunding. SAGE Open. https://doi.org/10.1177/2158244018805808

Frydrych D, Bock AJ, Kinder T (2016) Creating Project Legitimacy- The Role of Entrepreneurial Narrative in Reward-Based Crowdfunding. In J. Meric, & J. B. Maque, International Perspectives on Crowdfunding (pp. 99–128). https://doi.org/10.1108/978-1-78560-315-020151006

Frydrych D, Bock AJ, Kinder T, Koeck B (2014) Exploring entrepreneurial legitimacy in reward-based crowdfunding. Ventur Cap 16(3):247–269. https://doi.org/10.1080/13691066.2014.916512

Galvagno M, Dalli D (2014) Theory of value co-creation: A systematic literature review. Managing Service Quality 24(6):643—683. https://doi.org/10.1108/msq-09-2013-0187

Gangi F (2013) La finanza etica durante le crisi finanziarie del nuovo millennio. Modelli teorici ed evidenze empiriche. Guida editori

Gangi F, Daniele LM (2017) Remarkable Funders: How Early-Late Backers and Mentors Affect Reward-Based Crowdfunding Campaigns. International Business Research 10(11):58–69. https://doi.org/10.5539/ibr.v10n11p58

Gangi F, Daniele LM, Varrone N (2020). How do corporate environmental policy and corporate reputation affect risk-adjusted financial performance?. Bus Strategy Environ 29(5):1975–1991. https://doi.org/10.1002/bse.2482

Gangi F, Meles A, Monferrà S, Mustilli M (2020) Does corporate social responsibility help the survivorship of SMEs and large firms?. Glob Finance J 43:100402. https://doi.org/10.1016/j.gfj.2018.01.006

Gangi F, Méric J, Jardat R, Daniele LM (2019) Business for Society. Routledge

Gangi F, Mustilli M, Varrone N (2019) The impact of corporate social responsibility (CSR) knowledge on corporate financial performance: evidence from the European banking industry. J Knowl Manag 23(1):110–134. https://doi.org/10.1108/jkm-04-2018-0267

Gerber EM, Hui J (2013) Crowdfunding: Motivations and deterrents for participation. ACM Trans Comput-Hum Interact 20(6), Article 34. https://doi.org/10.1145/2530540

Gerber L, Hui J (2016) Crowdfunding: how and why people participate. International perspectives on crowdfunding, 37–64. https://doi.org/10.1108/978-1-78560-315-020151003

Giudici G, Rossi-Lamastra C (2018). Crowdfunding of SMEs and Startups: When Open Investing Follows Open Innovation. Researching Open Innovation in SMEs, World Scientific: 377–396. https://doi.org/10.1142/978 9813230972_0012

Greenwood M (2007) Stakeholder engagement: Beyond the myth of corporate responsibility. J Bus Ethics 74(4):315–327. https://doi.org/10.1007/s10551-007-9509-y

Grigore G, Molesworth M, Watkins R (2017) New corporate responsibilities in the digital economy. In Corporate social responsibility in the post-financial crisis era (pp. 41–62). Palgrave Macmillan, Cham. https://doi.org/10.1007/978-3-319-40096-9_3

Grönroos C (2011) Value co-creation in service logic: A critical analysis. Mark Theory 11(3):279–301. https://doi.org/10.1177/1470593111408177

Gummesson E (2006) Many-to-many marketing as grand theory. In R. F. Lusch, & S. L. Vargo (Eds.), The service-dominant logic of marketing, 339–353. New York: M.E. Sharpe

Gupta S, Lehman DR (2005) Managing Customers as Investments. Upper Saddle River, NJ: Wharton School Publishing

Hardy W (2013) How to perfectly discriminate in a crowd? A theoretical model of crowdfunding. University of Warsaw Faculty of Economic Sciences Working Papers 16(16):1–28

Hemer J (2011) A snapshot on crowdfunding (No. R2/2011). Arbeitspapiere Unternehmen und Region

Hennig-Thurau T, Gwinner KP, Walsh G, Gremler DD (2004) Electronic word-of-mouth via consumer-opinion platforms: What motivates consumers to articulate themselves on the Internet? J Interact Mark 18(1):38–52. https://doi.org/10.1002/dir.10073

Höchstädter AK, Scheck B (2015) What's in a name: An analysis of impact investing understandings by academics and practitioners. J Bus Ethics 132(2):449–475. https://doi.org/10.1007/s10551-014-2327-0

Hong L, Page SE (2004) Groups of diverse problem solvers can outperform groups of high-ability problem solvers. Proc Natl Acad Sci USA 101(46):16385–16389. https://doi.org/10.1073/pnas.0403723101

Hörisch J (2019) Take the money and run? Implementation and disclosure of environmentally-oriented crowdfunding projects. J Clean Prod 223:127–135. https://doi.org/10.1016/j.jclepro.2019.03.100

Hornuf L, Klus MF, Lohwasser TS, Schwienbacher A (2020) How do banks interact with fintech startups?. Small Bus Econ, 1–22. https://doi.org/10.1007/s11187-020-00359-3

Hornuf L, Schmitt M, Stenzhorn E (2018) Equity crowdfunding in Germany and the United Kingdom: Follow-up funding and firm failure. Corp Gov 26(5):331–354. https://doi.org/10.1111/corg.12260

Hornuf L, Schwienbacher A (2018) Market mechanisms and funding dynamics in equity crowdfunding. J Corp Finance 50:556–574. https://doi.org/10.1016/j.jcorpfin.2017.08.009

Howe J (2006) The rise of crowdsourcing. Wired magazine 14(6):1–4

Hu M, Li X, Shi M (2015) Product and pricing decisions in crowdfunding. Mark Sci 34(3):331–345. https://doi.org/10.1287/mksc.2014.0900

Illia L, Romenti S, Rodríguez-Cánovas B, Murtarelli G, Carroll CE (2017) Exploring corporations' dialogue about CSR in the digital era. J Bus Ethics 146(1):39–58. https://doi.org/10.1007/s10551-015-2924-6

Iyer R, Khwaja AI, Luttmer EF, Shue K (2009) Screening in new credit markets: Can individual lenders infer borrower creditworthiness in peer-to-peer lending?. AFA 2011 Denver meetings paper. https://doi.org/10.2139/ssrn.1570115

Kesavan R, Bernacchi MD, Mascarenhas OA (2013) Word of mouse: CSR communication and the social media. Manag Int Rev 9(1):58–66

Kim K, Viswanathan S (2016) The 'Experts' in the Crowd: The Role of 'Expert' Investors in a Crowdfunding Market. In TPRC 41: The 41st research conference on communication, information and internet policy

Kim K, Viswanathan S (2018) The 'Experts' in the Crowd: The Role of Experienced Investors in a Crowdfunding Market. To Appear in MIS Quarterly. https://doi.org/10.2139/ssrn.2258243

Kim MJ, Hall CM (2019). Can co-creation and crowdfunding types predict funder behavior? An extended model of goal-directed behavior. Sustainability 11(24):7061. https://doi.org/10.3390/su11247061

Kirby E, Worner S (2014). Crowd-funding: An infant industry growing fast. IOSCO Research Department

Koch JA, Siering M (2015) Crowdfunding Success Factors: The Characteristics of Successfully Funded Projects o Crowdfunding Platforms in: Proceedings of the 23rd European Conference on Information Systems, AIS, Münster, Germany

KPMG (2019) The Pulse of Fintech H2 2019. February 2020

Kunze C, Becker S (2014) Energy democracy in Europe. A survey and outlook. Rosa-LuxemburgStiftung Report. Brussels, July 2014

Kuppuswamy, V., & Bayus, B. L. (2018). Crowdfunding creative ideas: The dynamics of project backers. In The economics of crowdfunding (pp. 151–182). Palgrave Macmillan, Cham. https://doi.org/10.1007/978-3-319-66119-3_8

Lee I, Shin YJ (2018) Fintech: Ecosystem, business models, investment decisions, and challenges. Bus Horiz 61:35—46. https://doi.org/10.1016/j.bushor.2017.09.003

Lehner OM (2013) Crowdfunding social ventures: a model and research agenda. Ventur Cap 15(4):289–311. https://doi.org/10.1080/13691066.2013.782624

Lehner OM, Nicholls A (2014) Social finance and crowdfunding for social enterprises: A public–private case study providing legitimacy and leverage. Ventur Cap 16(3):271–286. https://doi.org/10.1080/13691066.2014.925305

Li Z, Duan JA (2016) Network externalities in collaborative consumption: Theory, experiment, and empirical investigation of crowdfunding. Working Paper

Lipusch N, Dellermann D, Bretschneider U, Ebel P, Leimeister JM (2020) Designing for Crowdfunding Co-creation - How to Leverage the Potential of Backers for Product Development. Bus Inf Syst Eng. https://doi.org/10.1007/s12599-019-00628-w

Löher J, Schneck S, Werner A (2018) A research note on entrepreneurs' financial commitment and crowdfunding success. Ventur Cap 20(3):309–322. https://doi.org/10.1080/13691066.2018.1480864

Lusch RF, Vargo SL (2014) Service-dominant logic: Premises, perspectives, possibilities. Cambridge University Press: Cambridge

Mair J, Martí I (2006) Social entrepreneurship research: A source of explanation, prediction, and delight. J World Bus 41(1):6–44. https://doi.org/10.1016/j.jwb.2005.09.002

Marom D, Robb A, Sade O (2014) Gender dynamics in crowdfunding (kickstarter). SSRN Work. Pap, 2442954

Marom D, Swart R, Grell KB (2015) Crowdfunding: The corporate era. Elliott and Thompson Ltd

Marom S (2017). Social responsibility and crowdfunding businesses: a measurement development study. Soc Responsib J 13(2):235–249. https://doi.org/10.1108/srj-07-2016-0118

Mastrangelo L, Cruz-Ros S, Miquel-Romero MJ (2019) Crowdfunding success: the role of co-creation, feedback, and corporate social responsibility. Int J Entrepreneurial Behav Res 26(3):449–466. https://doi.org/10.1108/ijebr-06-2019-0391

McGuire E (2017) Can Equity Crowdfunding Mitigate the Gender Gap in Entrepreneurship? In 2017 ASSA Annual Meeting, American Economic Association, Chicago, IL

Méric J (2016) Crowdfunding in Present Society: Deconstructing the Zeitgeist. In: Méric J, Maque I, Brabet J (eds) International Perspectives on Crowdfunding, Emerald Group Publishing Limited, pp 201–218. https://doi.org/10.1108/978-1-78560-315-020151012

Miglo A, Miglo V (2019) Market imperfections and crowdfunding. Small Bus Econ 53(1):51–79. https://doi.org/10.1007/s11187-018-0037-1

Mochkabadi K, Volkmann CK (2020) Equity crowdfunding: a systematic review of the literature. Small Bus Econ 54:75–118. https://doi.org/10.1007/s11187-018-0081-x

Mollick ER (2014) The dynamics of crowdfunding: An exploratory study. J Bus Ventur 29(1):1–16. https://doi.org/10.1016/j.jbusvent.2013.06.005

Mollick ER (2016) Containing multitudes: the many impacts of Kickstarter funding. Working Paper. https://doi.org/10.2139/ssrn.2808000

Mollick ER, Kuppuswamy V (2014) After the Campaign: Outcomes of Crowdfunding (January 9, 2014). UNC Kenan-Flagler Research Paper No. 2376997. https://doi.org/10.2139/ssrn.2376997

Mollick ER, Nanda R (2016) Wisdom or Madness? Comparing crowds with expert evaluation in funding the arts. Manage Sci, 62(6):1533–155. https://doi.org/10.1287/mnsc.2015.2207

Moritz A, Block JH (2016) Crowdfunding: A literature review and research directions. In Crowdfunding in Europe (pp. 25–53). Springer, Cham. https://doi.org/10.1007/978-3-319-18017-5_3

Nahapiet J, Ghoshal S (1998, April 1) Social capital, intellectual capital, and the organizational advantage. Acad Manag Rev 23(2):242–66. https://doi.org/10.5465/amr.1998.533225

Nevin S, Gleasure R, O'Reilly P, Feller J, Li S, Cristoforo J (2017) Social Identity and Social Media Activities in Equity Crowdfunding. Proceedings of the 13th International Symposium on Open Collaboration 11:1–8. https://doi.org/10.1145/3125433.3125461

Nitani M, Riding A (2017) On Crowdfunding Success: Firm and Owner Attributes and Social Networking. Emerging Trends in Entrepreneurial Finance Conference. https://doi.org/10.2139/ssrn.2945081

Nitani M, Riding A, He B (2019, July 3) On Equity Crowdfunding: Investor Rationality and Success Factors. Venture Capital 21(2–3):243–272. https://doi.org/10.1080/13691066.2018.1468542

OECD (2019) Financing SMEs and entrepreneurs. An OECD scoreboard

Ordanini A, Miceli L, Pizzetti M, Parasuraman A (2011) Crowd-funding: Transforming customers into investors through innovative service platforms. J Serv Manag 22(4):443–470. https://doi.org/10.1108/09564231111155079

Payne AF, Storbacka K, Frow P (2008) Managing the co-creation of value. J Acad Mark Sci 36(1):83–96. https://doi.org/10.1007/s11747-007-0070-0

Paschen J (2017) Choose wisely: Crowdfunding through the stages of the startup life cycle. Bus Horiz 60(2):179–188. https://doi.org/10.1016/j.bushor.2016.11.003

Pérez A, Del Bosque IR (2013) Measuring CSR image: three studies to develop and to validate a reliable measurement tool. J Bus Ethics 118(2):265–286. https://doi.org/10.1007/s10551-012-1588-8

Perrini F, Vurro C (2010) Corporate sustainability, intangible assets accumulation and competitive advantage constraints. Symphonya. Emerging Issues in Management (2):25–38. https://doi.org/10.4468/2010.2.03perrini.vurro

Petersen MA (2004) Information: Hard and soft (Working paper, Northwestern University). Retrieved from http://www.kellogg.northwestern.edu/faculty/petersen/htm/papers/softhard.pdf

Pierrakis Y (2019) Peer-to-peer lending to businesses: Investors' characteristics, investment criteria and motivation. Int J Entrepreneurship Innov 20(4):239–251. https://doi.org/10.1177/1465750319842528

Piva E, Rossi-Lamastra C (2018) Human capital signals and entrepreneurs' success in equity crowdfunding. Small Bus Econ 51:667–686. https://doi.org/10.1007/s11187-017-9950-y

Prahalad CK, Ramaswamy V (2004a) Co-creation experiences: The next practice in value creation. J Interact Mark 18(3):5–14. https://doi.org/10.1002/dir.20015

Prahalad CK, Ramaswamy V (2004b) Co-creating unique value with customers. Strategy & leadership

Puschmann T (2017) Fintech. Bus Inf Syst Eng 59:69–76. https://doi.org/10.1007/s12599-017-0464-6

Quero MJ, Ventura R (2019). Value proposition as a framework for value co-creation in crowdfunding ecosystems. Mark Theory 19(1):47–63. https://doi.org/10.1177/1470593118772213

Ralcheva A, Roosenboom P (2016) On the road to success in equity crowdfunding. SSRN Working Paper, No. 2727742. https://doi.org/10.2139/ssrn.2727742

Ramos IM (2018) El crowdfunding inmobiliario,¿ colectivización de la propiedad?. El notario del siglo XXI: revista del Colegio Notarial de Madrid (79):25–27

Ranjan KR, Read S (2016) Value co-creation: concept and measurement. J Acad Mark Sci 44:290–315. https://doi.org/10.1007/s11747-014-0397-2

Ridley-Duff R (2009) Co-operative social enterprises: company rules, access to finance and management practice. Soc Enterp J 5(1):50–68. https://doi.org/10.1108/17508610910956408

Rindova VP, Petkova AP, Kotha S (2007) Standing out: How new firms in emerging markets build reputation. Strateg Organ 5(1):31–70. https://doi.org/10.1177/1476127006074389

Russo-Spena T, Mele C (2012) "Five Co-s" in innovating: a practice-based view. J Serv Manag 23(4):527–55. https://doi.org/10.1108/09564231211260404

Scholtens B (2009) Corporate social responsibility in the international banking industry. J Bus Ethics 86(2):159–175. https://doi.org/10.1007/s10551-008-9841-x

Schumpeter J (1934) The theory of economic development Harvard University Press. Cambridge, MA

Schumpeter J (1964) Business cycles: A theoretical, historical and statistical analysis of the capitalist process. Abridged, with an introd., by Rendigs Fels. McGraw-Hill

Schwienbacher A, Larralde B (2010) Crowdfunding of small entrepreneurial ventures. HANDBOOK OF ENTREPRENEURIAL FINANCE, Oxford University Press, Forthcoming. https://doi.org/10.2139/ssrn.1699183

Shafi K (2019) Investors' evaluation criteria in equity crowdfunding. Small Bus Econ. https://doi.org/10.1007/s11187-019-00227-9

Simonsohn U, Ariely D (2008) "When Rational Sellers Face Nonrational Buyers: Evidence from Herding on eBay". Manage Sci 54(9):1624–1637. https://doi.org/10.1287/mnsc.1080.0881

Skirnevskiy V, Bendig D, Brettel M (2017) The Influence of Internal Social Capital on Serial Creators' Success in Crowdfunding. Entrep Theory Pract 41(2), 209–236. https://doi.org/10.1111/etap.12272

Spanos L (2018) Complementarity and interconnection between CSR and crowdfunding: A case study in Greece. In Corporate responsibility and digital communities (pp. 29–49). Palgrave Macmillan, Cham. https://doi.org/10.1007/978-3-319-63480-7_3

Spence M (1973) Job Market Signaling. Q J Econ 87(3):355–374. https://doi.org/10.2307/1882010

Stanko MA, Henard DH (2017) Toward a better understanding of crowdfunding, openness and the consequences for innovation. Res Policy 46(4):784–798. https://doi.org/10.1016/j.respol.2017.02.003

Starteed (2019) Il crowdfunding in Italia

Stein JC (2002) Information production and capital allocation: Decentralized versus hierarchical firms. J Finance 57(5):1891–1921. https://doi.org/10.1111/0022-1082.00483

Stevenson RM, Kuratko DF, Eutsler J (2019) Unleashing main street entrepreneurship: Crowdfunding, venture capital, and the democratization of new venture investments. Small Bus Econ 52:375–393. https://doi.org/10.1007/s11187-018-0097-2

Surowiecki J (2004) The Wisdom of Crowds: Why the Many are Smarter than the Few and How Collective Wisdom Shapes Business, Economies, Societies, and Nations. New York: Doubleday

Surowiecki J (2005) The wisdom of crowds. New York: Anchor Books

Surroca J, Tribó JA, Waddock S (2010) Corporate responsibility and financial performance: the role of intangible resources. Strateg Manag J 31(5):463–490. https://doi.org/10.1002/smj.820

Tang Z, Hull CE, Rothenberg S (2012) How corporate social responsibility engagement strategy moderates the CSR–financial performance relationship. J Manage Studies 49(7):1274–1303. https://doi.org/10.1111/j.1467-6486.2012.01068.x

Testa S, Roma P, Vasi M, Cincotti S (2020) Crowdfunding as a tool to support sustainability-oriented initiatives: Preliminary insights into the role of product/service attributes. Bus Strategy Environ 29(2):530–546. https://doi.org/10.1002/bse.2385

Thies F, Wessel M (2014). The Circular Effects of Popularity Information and Electronic Word-of-Mouth on Consumer Decision-Making: Evidence from a Crowdfunding Platform. Paper presented at the European Conference on Information Systems, Tel Aviv, Israel

Tomboc GFB (2013) The lemons problem in crowdfunding. J. Marshall J. Info. Tech. & Privacy L

UN (2020) The Sustainable Development Goals Report 2020

Van Marrewijk M (2003) Concepts and definitions of CSR and corporate sustainability: Between agency and communion. J Bus Ethics 44(2–3):95-105. https://doi.org/10.1023/a:1023331212247

Vargo SL, Lusch RF (2004) Evolving to a New Dominant Logic for Marketing. J Mark 68(1):1–17. https://doi.org/10.1509/jmkg.68.1.1.24036

Vargo SL, Maglio PP, Akaka MA (2008) On value and value co-creation: A service systems and service logic perspective. Eur Manag J 26(3):145–152. https://doi.org/10.1016/j.emj.2008.04.003

Vismara S (2016) Equity retention and social network theory in equity crowdfunding. Small Bus Econ 46(4):579–590. https://doi.org/10.1007/s11187-016-9710-4

Vismara S (2018) Information cascades among investors in equity crowdfunding. Entrep Theory Pract 42(3):467–497. https://doi.org/10.1111/etap.12261

Vismara S (2019) Sustainability in equity crowdfunding. Technol Forecast Soc Change 141:98–106. https://doi.org/10.1016/j.techfore.2018.07.014

Vulkan N, Åstebro T, Sierra MF (2016, June 1) Equity crowdfunding: A new phenomena. J Bus Ventur Insights 5:37–49. https://doi.org/10.1016/j.jbvi.2016.02.001

Wald A, Holmesland M, Efrat K (2019) It Is Not All About Money: Obtaining Additional Benefits Through Equity Crowdfunding. J Entrepreunership 28(2):270–294. https://doi.org/10.1177/0971355719851899

Wanderley LSO, Lucian R, Farache F, de Sousa Filho JM (2008) CSR information disclosure on the web: a context-based approach analysing the influence of country of origin and industry sector. J Bus Ethics 82(2):369–378. https://doi.org/10.1007/s10551-008-9892-z

Wehnert P, Baccarella CV, Beckmann M (2019) In crowdfunding we trust? Investigating crowdfunding success as a signal for enhancing trust in sustainable product features. Technol Forecast Soc Change 141:128–137. https://doi.org/10.1016/j.techfore.2018.06.036

Zhu F, Zhang X (2010) Impact of online consumer reviews on sales: The moderating role of product and consumer characteristics. J Mark 74(2):133–148. https://doi.org/10.1509/jm.74.2.133

April 2005 and October 2007, Facebook received three series of VC financing from Accel Partners (Series A), Greylock Partners and Meritech Capital (Series B) and Microsoft (Series C). The implied evaluation for these rounds was around $12.7 million, $500 million and $15 billion, respectively. On May 2012, Facebook raised $16 billion through an Initial Public Offering (IPO), which gave it a market value of $102.4 billion which is approximately 8000 times the implied value received during the first VC financing stage. Other well-known VC-backed companies include Apple, Google and Twitter in the US, and Skype, Zalando and JustEat in Europe.

One characteristic that VCs have in common is that they are not agnostic about the target industry. When VC investors pick portfolio firms, they usually pay considerable attention to the industry in which these firms operate and naturally prefer sectors with huge growth potential.[1]

VC investments are usually classified into three categories, as follows:

1. **Seed stage.** Financing provided to enable a business idea to get off the ground, to elaborate a business plan, conduct additional research and realise the product prototype, prior to large-scale manufacturing.
2. **Startup stage**. Financing to develop a firm's products and fund its initial marketing. Firms may be in the process of being formed, or they may have been in business for a short time without trading, therefore generating no revenues.
3. **Later stage**. Financing provided for the expansion of an operating firm, which may or may not have reached its break-even point.

[1] Don Valentine, founder of Sequoia Capital, one of the most successful US VC firms, provided an in-depth explanation, during an inspiring speech given at Stanford University, of the main criteria used to target firms. Among the points made was this: "I hammered out this more intuitive investment selection process based on huge markets and solutions that made a significant short-term commercial sense. Six hundred investments, thirty-odd years later, we're still using basically the same selection criteria in choosing companies that have huge, huge markets. It turns out it's much easier to build a new startup company in an environment where the markets are large than it is to try to develop a market based on some technology for which there's not an obvious solution. And that's been sort of the approach we've continued with which I stared somewhere about 1967 or so." The full speech can be found on YouTube: https://www.youtube.com/watch?v=nKN-abRJMEw.

Later-stage VC investors are likely to invest in firms already backed by other VC firms.

VC firms vary considerably according to their areas of specialisation (in terms of industry, geography and investment stage), their source of funds, and their management and governance structures. Traditionally, two types of VC firms are identified. **Captive VC firms** are those in which one shareholder provides the capital, i.e. where the parent organisation allocates money to the VC firm from its own internal sources. These firms are typically subsidiaries of financial institutions such as banks, but they can also be owned and operated by other entities, including non-financial companies. One famous example of a captive VC firm is Alibaba Group Partners, the investment arm of Alibaba group, which provides venture funding to small companies who wish to take advantage of the resources offered by Alibaba's ecosystem. In contrast, **independent VCs** are those in which outside investors provide the majority of the capital. Most VC firms are structured in this way, with varying numbers of shareholders (from a few units to several thousand) among high-net-worth individuals, financial institutions, non-financial companies and nonprofit organisations. Examples of independent VC firms include, in Europe, HV Holtzbrinck Ventures and United Ventures, and, in the US, Sequoia Capital.

Regardless of the origin of the funds, the most common form of organisation of VC firms worldwide is the VC fund (Cumming and Johan 2013). It is typically organised under a Limited Partners model, with a structure involving a number of entities, as illustrated in Fig. 5.1.

The General Partner (GPs) are the main players within a VC fund, with the legal power to raise money from the investors (Limited Partners [LPs]) and invest it in a number of portfolio companies (usually between 10 and 20 investee firms) with the purpose of creating value for the LPs. More specifically, the GP appoints an investment team with strong expertise in the target sectors or geographies of the fund and sometimes also an investment adviser firm (often affiliated with the GP).

Furthermore, the VC fund business model can be subdivided into the following four main phases.

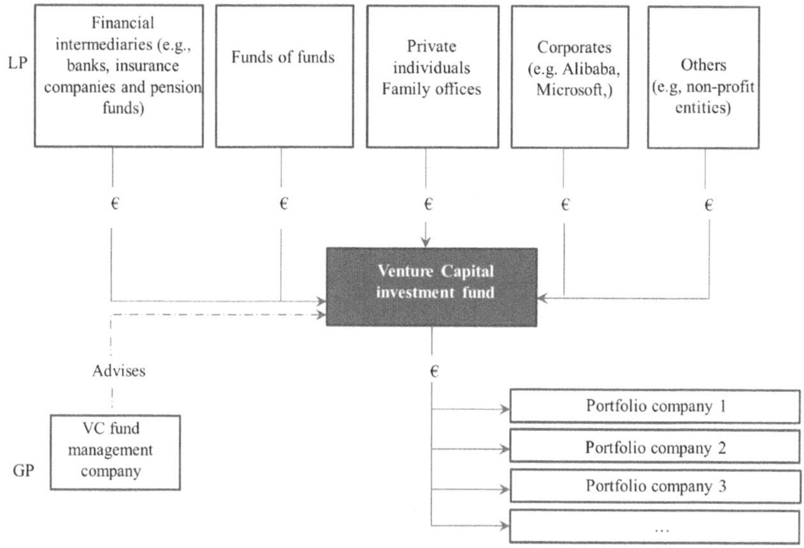

Fig. 5.1 Main entities involved in a VC fund (*Source* Adapted from EVCA)

- *Fundraising and fund closing*

Once approval by the regulatory agency has been obtained, the GP launches its VC fund to collect money from LPs. Except for captive VC firms, which gather capital mainly from their parent companies, GPs generally raise funds from outside investors, including, among others, nonprofit organisations and financial intermediaries. A first closing of the VC fund takes place when the GP has reached its fundraising target. From that point, LPs may subsequently be asked to make the first capital injections to cover organisational expenses, pay the GPs' fees and conclude initial deals. After the first closing, subsequent closings are typically held in approximately twelve-month intervals (the so-called marketing period).

- *Investment period (or commitment period)*

After the fund closes, the investment period starts, the length of which varies depending on the VC fund's investment strategy. Many have an investment period of five years with an option for the GP to extend for one or two additional years. During the investment period, VC firms search intently for high-quality businesses to invest in according to the strategy proposed to the LPs and before the investment period expires. As widely documented by academic research (Chemmanur et al. 2011; Megginson et al. 2019), much of VC firms' success derives from their intensive screening and selection process to pick "winning firms" with favourable future business prospects. It is not by chance that VC financing is referred to a "black swan" phenomenon. In fact, only a small minority of all firms (in the US about 1%) obtain funds from VCs (Lerner and Tag 2013).

- *Holding period*

Once the investment period has expired, VC funds can hold equity stakes in portfolio companies for an additional period (commonly five to seven years) in order to maximise their value from the firms. This period can be extended for a few years by mutual agreement, but there are various financial incentives in place for VC funds to return money to the investors beforehand. Thus, in some cases, a fund might effect an opportunistic early exit, but in other cases the GP might extend the holding period due to low performance of several portfolio companies.

During the holding period, VCs will not simply monitor their investments passively, as commercial banks usually do with their borrower firms. VC firms' success—and the fact that their investee companies are acknowledged to perform better than comparable firms (Chemmanur et al. 2011)—is not just a result of its superior scouting and screening. The broad consensus is that another reason is because, during the holding period, they add value to their portfolio firms by providing assistance in defining strategic planning, in management recruitment and compensation, effective monitoring, access to their network of contacts (i.e. banks, suppliers and customers) and expertise in operational planning. Such value-adding services are especially welcome in early-stage and high-tech businesses. These value-added services mean that VC-backed firms' performance is likely to be better than that of non-VC-backed firms, even if they were on an even footing at the investment date. The evidence for this is even more compelling for enterprises backed by highly reputable VC firms which have a more active post-investment involvement in the corporate governance of their portfolio firms (Krishnan et al. 2011).

- *Divestment period and redistribution*

Equity stakes held by VC funds are typically not liquidated all at once but in separate liquidity events as and when directed by the investment team, primarily during a divestment period ending. At the end of the fund's term (typically four to six years after the investment period), its remaining investments must be liquidated with the proceeds distributed to investors in accordance with the distribution waterfall.

The exit is the process by which VC firms realise a return on their investment after a certain amount of time, so it is therefore a natural part of the life cycle of every VC transaction. Also, the number of successful exits a VC firm achieves has a strong influence on its ability to attract investors and raise funds. Accordingly, the potential exit opportunities offered by an investment represent a highly important factor in an investor's decision about whether or not to invest in a company. Hence, exits are a key focus of attention from the earliest stages of a deal.

VC firms can use different channels to liquidate their stakes in portfolio companies. Below is a brief analysis of the most important and widely used exit strategies:

- **Initial Public Offering (IPO)**. Under an IPO, a firm's shares are listed on the stock market for the first time, so the VC firm will be able to sell its stake to public investors. This is one of the best exit strategies for a VC firm, because, if market conditions are particularly favourable, it can generate the highest return on the investment. Moreover, a large part of a VC's reputation derives from its ability to bring firms it has backed public, and, since reputation is critical for future fundraising, VCs have an incentive to do this.
- **Trade sale**. A trade sale occurs when a VC firm sells its equity stake to a third party, often an industry peer or a competitor. This exit route can be advantageous for a VC because it provides a complete and immediate exit from the investment. Furthermore, the investor can get a higher value for the company compared to other exit routes, as strategic bidders might be willing to pay a premium due to potential synergies.
- **Secondary sale**. In a secondary sale, a VC firm's stake is sold to another VC (or PE investor). VCs usually choose this exit strategy

when they are not willing to continue to invest in a firm, which, in turn, is not yet ready for a trade sale or an IPO.

- **Buy-back**. In a buy-back, the VC investor sells its stake back to the entrepreneur or company from which the shares were purchased. A buy-back is usually proposed by an entrepreneur who does not wish to go public or sell his/her stake.
- **Write-off**. A write-off occurs when a VC investor walks away from its investment, i.e. when it is clear the expected return will be impossible to realise, often because the investee firm has become insolvent.

5.3 ESG Implementation in the VC Industry: Motivation, Diffusion and Strategies

In recent decades, ESG issues have been attracting increasing investor interest across industries. Accordingly, there are now more asset management companies implementing ESG criteria in their investment decision processes. Nonetheless, there remains a perception among some market participants that the adoption of ESG principles by the VC industry is negligible compared to other financial areas such as banking and mutual funds. From a theoretical perspective, this idea is supported by two main arguments. First, a number of VCs, particularly those with substantial investments in areas notorious for being "high impact" (biotech, green-tech, med-tech and reg-tech) may not even recognise the need to think through ESG guidelines in their decision-making process. A noteworthy example is Obvious Ventures, a VC firm co-founded by Medium CEO and Twitter co-founder Evan Williams. Obvious Ventures' mission is to provide solutions to the biggest problems faced by humanity while financing the most valuable companies, yet its partners reject the "impact investor" label. Aiming to change the world in a positive way is in fact a cornerstone of its corporate culture.[2] Second, VCs tend to invest in tech firms that have a good environmental reputation, which is largely facilitated by the fact that these companies produce data and services rather than goods and materials and so do not consume large quantities of natural resources or generate excessive emissions.

[2] "Obvious Ventures aims to surpass impact investing model". *Wall Street Journal*, March 18, 2019.

Despite the above arguments, VC is an intriguing case from an ESG perspective since there are multiple reasons why VC firms would have to deal with significant governance and social risks. Without claiming to be exhaustive, we aim to shed light on some of these critical issues.

A first important governance issue is that founders of VC-backed firms maintain very strong control over the company even after the IPO and the exit of the VC. For instance, when Facebook went public in 2012, it offered a dual-class stock, which gave Zuckerberg about 60% of shareholder votes while controlling less than a fifth of the company's shares. Among the many recent VC-backed IPOs, the adoption of dual-class shareholder structures has not been uncommon, raising the issue of limited controlling rights for public investors. This practice—which might find justification because of the fundamental contribution that founders can continue to make to the growth of the business—risks creating conflict between existing and new shareholders. In such a context, VCs need to encourage their portfolio companies to adopt strong corporate governance mechanisms, to avoid an excess of founders' power hindering the value creation process and damaging the VC firm's reputation.

A second issue for VCs, which was touched on above, is that companies founded by women receive a tiny fraction of the sum total of VC funds.[3] Balachandra (2020) argues that the substantial gender imbalance in VC investing may derive from the massive disparity between the numbers of male and female VC investors. Specifically, she argues that, with 93% of VCs being men, this may cause "gender bias that prevents women from having their ventures being considered for funding from the pitch as well as from obtaining opportunities to pitch VCs in consideration for funding due to biases in the evaluations of the businesses themselves" (p. 261).

Other social risks to which VCs are exposed are directly connected to the core business issues typical of many VC-backed companies. The first is the lack of diversity in the tech industry; not only is this socially unacceptable but it presents significant opportunity costs for the companies concerned. The gender and racial bias evident in artificial intelligence systems sold by both big and small tech companies is emblematic of

[3] We investigate this issue more thoroughly in Sect. 5.6.

this problem.[4] The second issue arises because of the large amounts of personal information handled by several VC-backed tech firms, which must address the privacy of customer data and the ethical use thereof.

Another pressing social issue for most VC-backed firms is labour and working conditions. Indeed, early-stage firms tend not to have a human resources policy, nor have they developed the skills to manage employment relationships effectively. Moreover, this deficiency, which may have limited consequences to begin with because of the small number of workers involved, can become more problematic as a firm grows and starts managing more complex working environments.

With a growing awareness that ESG principles are just as fundamental to the VC industry, the following questions arise: How common are ESG criteria in the VC industry? What are the main factors that encourage VCs to engage in ESG actions? Which are the most widely used ESG investment strategies? What are the main characteristics of those VC firms that incorporate ESG guidelines into their decision-making process? Do VC firms report any significant relationship between their ESG engagement and returns on investment?

While academic research is yet to answer these key questions, a recent working paper (Botsari and Lang 2020) sponsored by European Investment Fund (EIF) has tried to provide some preliminary evidence on the integration of ESG in the European VC industry. The findings, which are derived from the EIF VC Survey 2019,[5] debunk the idea that compliance with ESG guidelines is less relevant to the VC industry than to other finance sectors. It seems that approximately 70% of VCs are already incorporating ESG guidelines into their investment decision processes. The reasons for this strategic choice are summarised in Fig. 5.2, which shows that VC firms who engage in ESG actions see the pursuit of ethical or strategic purpose as more important than financial or regulatory compliance.

Three out of four VC firms state that they adopt ESG strategies out of consideration for ethical or social responsibility, while about 50% of ESG-compliant VC firms are motivated by a desire to inspire a transformation towards responsible business practices in their portfolio firms

[4] One famous example is the "Tay" chatbot launched by Microsoft to connect with millennials, but which started tweeting racist and anti-feminist comments.

[5] The EIF VC Survey 2019 is "a survey among VC general partner (GP)/management companies headquartered in the EU27, the UK and other European countries".

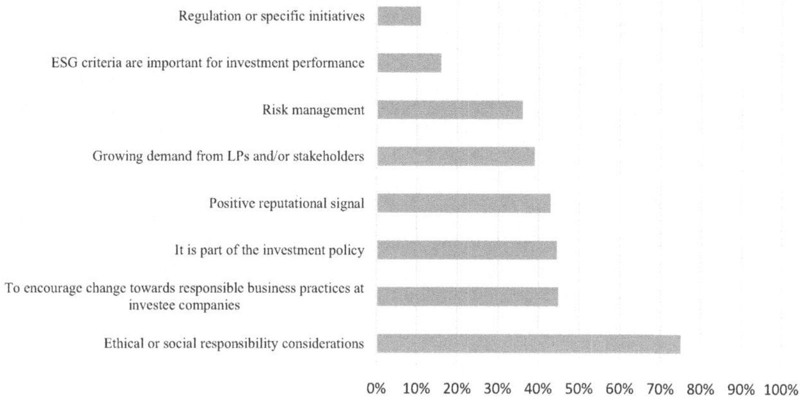

Fig. 5.2 European VC funds' motivations for ESG engagement (*Source* Adapted from Botsari and Lang [2020])

and to integrate ESG principles into their investment policy. More than one in three VC fund managers, on the other hand, are motivated by strategic factors, such as the reputational benefit associated with ESG investing and the growing demand from LPs. Surprisingly, about one in three VC investors implement ESG guidelines in their decision-making as a means of pursuing financial goals, such as performance improvement or risk mitigation, while only one in ten VC firms cite regulatory initiatives as a significant factor determining an ESG orientation.

Interestingly, among those VC firms with an ESG policy, ESG information mainly serves as a tool to support the selection of investee firms (see Fig. 5.3). "Negative screening" techniques are used by 50% of VCs to exclude companies belonging to non-ESG-friendly industries, including fossil fuels, alcohol, tobacco, gambling, firearms and military. This is a simple approach that guarantees a limited exposure to controversial industries but fails to exclude firms with poor ESG performance that operate in generally ESG-friendly sectors. Just less than 40% of VCs adopt positive screening tools to identify firms with high ESG performance. Less widespread among VCs are the kinds of ESG strategy that require the greater involvement of GPs, such as active monitoring of investee companies, or the full integration of ESG criteria into the evaluation of investment opportunities.

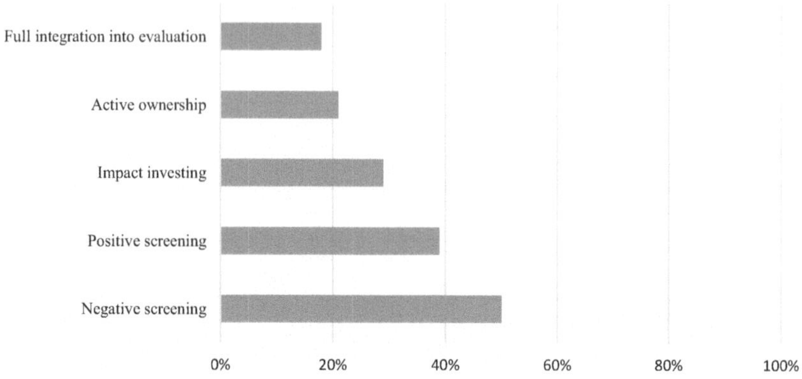

Fig. 5.3 ESG strategies of European VC funds (*Source of data* Botsari and Lang [2020])

Botsari and Lang (2020) analysed to what extent the level of ESG engagement, the ESG drivers and the investment strategies vary according to VC firms' characteristics, like location, size, investment stage focus and gender diversity in the management team. They report several interesting findings. First, investment activities incorporating ESG criteria are particularly prominent in the Nordic countries, the UK, Ireland and France, and less widespread in Germany, Austria, Switzerland and the Central, Eastern and South-eastern European (CESEE) countries. This evidence is consistent with the institutional theory (Baldini et al. 2018) which argues that country-level factors related to the social context in which a firm operates may either reduce or enhance ESG engagement and disclosure levels. Among the main studies, Ioannou and Serafeim (2012) find that significant cross-country ESG disclosure variations can be attributed to idiosyncratic institutional, cultural and political factors.

The second interesting result reported by Botsari and Lang (2020) is that the degree of ESG integration increases with a VC firm's reputation and also increases as it moves from early-stage to later-stage financing. The positive correlation with reputation is probably due to VCs' need to preserve their fundraising capabilities, which they are increasingly being pressured by LPs to do. Also, public-domain ESG information about an investee firm will increase relative to its age, facilitating the integration of ESG standards into the investment process at a later stage. Third, the authors observe a positive relationship between gender diversity in a VC firm's management team and the level of ESG involvement. Even though these findings are weak from a methodological point of view, they are of interest because they once more call the VC industry's gender gap into question.

Botsari and Lang's study (2020) also contributes to the discussion about the effects of VCs' ESG engagement on their investment returns, whether positive, neutral or negative. It transpires that VC fund managers concur that meeting ESG criteria can boost returns on investment and therefore allow them to "do good while doing well". This perception is consistent with a number of empirical studies providing evidence that engaging in ESG actions and maximising shareholder value are not mutually exclusive goals (*inter alia*, Awaysheh et al. 2020; Gangi et al. 2020). However, according to VC fund managers' experience, different ESG investment strategies vary in their degree of positive impact on returns on investment, and depending also on the level of involvement by the VC firm. It is no surprise to find that negative screening is considered the least effective strategy in terms of investment returns, despite being the most widely used. As explained above, this strategy is driven more by risk mitigation rather than by the maximisation of investment performance.

5.4 How Does ESG Compliance Contribute to VCs' Success?

In previous sections, we described the main phases of VC firms' activities and their approach to ESG policies. In this section, we aim to explain in more depth why ESG compliance can help VC firms raise funds and perform better in each of their investing phases; we will also provide some operational indications.

5.4.1 ESG and VC Firms' Fundraising

As we have seen, VC firms do not generally invest their own funds but rather convey financial resources from a plurality of investors to talented firms. As such, the first basic task of VC activity is to raise funds from potential LPs, such as pension funds, investment banks, insurance companies and funds of funds. Previous studies (Gompers and Lerner 1999) show that VC fundraising is influenced by a number of macroeconomic factors. For example, when the economy is growing fast, the climate is right for entrepreneurs to launch new companies and thus the demand for VC funds increases sharply. Macroeconomic factors can also affect the supply of VC funds: if interest rates rise, for instance, willingness of investors to subscribe to VC fund shares may fall.

Various authors (Gompers and Lerner 1999; Kaplan and Schoar 2005; Balboa and Marti 2007) demonstrate that VC fundraising, in addition to market factors, is determined by an individual VC firm's characteristics, such as its track record in terms of annual performance and number of VC-backed firms taken to an IPO. Although no study has as yet examined the causal relationship between fundraising power and ESG implementation within the VC industry, there are several reasons to believe that this relationship has been solidifying in recent years. First, a growing number of LPs, including pension funds, are implementing a more structured approach in handling ESG risks and opportunities. Regulatory changes seem to be the main factor driving the acceleration of ESG investing. According to a 2019 study conducted by State Street Global Advisors, which surveyed over 300 financial institutions, over half of European institutional investors said that compliance with regulation is their main reason for adopting ESG guidelines. This should come as no surprise considering that the Sustainability-Related Disclosure Regulation (SRDR) will come into force in March 2021.[6] The SRDR sets out detailed rules requiring fund managers to disclose how sustainability risks (environmental, social or governance) are incorporated into their investment decision-making processes and how these decisions impact on sustainability factors (a so-called adverse impacts statement). Other key reasons for LPs' adoption of ESG standards are avoidance of reputational risk, mitigation of ESG risks and also pursuit of financial benefits, such as reducing portfolio volatility risk and generating higher returns.

As a result of increasing demand for ESG by institutional investors, potential LPs might be well advised, during the fundraising phase, to view a copy of the VC fund's ESG policy as part of their investment due diligence. In this regard, the Principles for Responsible Investment (PRI) recommends that LPs submit the Responsible Investment Due Diligence Questionnaire (RIDDQ)[7] to GPs, the aim of which is to properly evaluate the integration of ESG standards into their investment practices.

[6] This set of rules is part of a legislative package proposed by the European Commission to encourage sustainable investment within the EU financial system.

[7] PRI (2015), "The Limited Partners' Responsible Investment Due Diligence Questionnaire".

The initiative is meant to encourage LPs to adopt an effective ESG due diligence approach and, ultimately, to stimulate more effective disclosure from GPs. Practical guidance for responsible investors has also been made publicly available by CDC, the UK government's development finance institution. Table 5.1 sets out the list of questions that CDC recommends

Table 5.1 Questions to aid in the assessment of a fund manager's ESG management system

Policies and processes

1. **Policy**. Are there formal policies and systems to manage ESG?
2. **Identifying opportunities**. Does the fund manager proactively identify opportunities for ESG improvements for its investments?
3. **Risks**. Are investments formally assessed for ESG risk and given an appropriate level of monitoring based upon this risk taking?
4. **Critical risks**. Can ESG considerations act as a block on a potential investment opportunity?
5. **Action plans**. Are formal action plans drawn up to address ESG deficiencies?
6. **Monitoring**. Are there any processes in place to manage/monitor ESG risks and to implement action plans for improvement?

Roles and responsibilities

7. **ESG resources.** Does the fund manager have a designated ESG professional?
8. **Top-level responsibility.** Has ESG responsibility been established at all levels including the fund's investment committee and governing body?
9. **Specialists.** Are specialist consultants/external technical experts used to assess and monitor high-risk investments? If so, when and who?
10. **Training.** Does the fund manager provide ESG training for all relevant staff?

ESG performance management

11. **Performance indicators**. Are key performance indicators in place to measure and track ESG performance at portfolio companies?
12. **Serious incidents.** Is there an established protocol for how to follow up serious incidents involving portfolio companies?

Reporting

13. **Communication lines.** Are defined lines of communication in place to report on ESG matters, to the fund manager's investment committee and governing body?
14. **Reporting to the fund's investors.** Are ESG matters reported on to the fund's investors at least annually and are any serious incidents reported immediately?

Stakeholder management

15. **Disclosure.** Does the fund manager publicly communicate on ESG matters through its annual report and website?
16. **Media.** Does the fund manager maintain a good relationship with the media on ESG aspects of their investments?

Source CDC (2018), *Toolkit on ESG for Fund Managers*

LPs ask GPs before making a decision about investing in a VC fund. Of course, this questionnaire is also useful for GPs in assessing the quality of their ESG management systems.

Despite the foregoing, from a VC firm's perspective implementing ESG guidelines is not merely a way of meeting LPs' expectations; it may also represent a means of capturing business opportunities and facing off the increased competition that has arisen in the VC industry in recent decades. In fact, an unprecedented number of VC firms worldwide are competing both for deals and for LPs' funds. According to Pitchbook (2020a), in the period 2009–2012 an average of 157 US VC funds were launched per year; in the four-year period 2016–2019, that figure rose to 289. In Europe, too, the numbers of new VC funds have been growing over the last ten years, albeit less dramatically (Pitchbook 2020b). As such, with VC markets becoming so highly competitive, differentiation has never been more necessary; to this end, ESG can be an effective strategy. This assertion is also consistent with previous literature analysing the effects of ESG policy on non-financial firms' outcomes. In fact, numerous studies have reported robust evidence that ESG can form a key component of a differentiation strategy by generating trust among consumers and fostering innovation (Pivato et al. 2008; Siegel and Vitaliano 2007; Porter and Kramer 2006). A strategic view of ESG for VC firms is one in which—in competitive environments—GPs improve their ESG performance as a means of differentiation and to avoid losing investors to more responsible competitors.

5.4.2 Role of ESG in Mitigating Problems of Information Asymmetry Faced by VC Firms

Existent literature (*inter alia*, Trester 1998; Cumming and Johan 2008) recognises the importance of information asymmetry in the VC industry. Several studies argue that VC firms differ markedly from other financial intermediaries to the extent that they invest in contexts where information asymmetry is more extreme. More specifically, the finance literature has made a distinction between *ex ante* and *ex post* information asymmetry. The former occurs because entrepreneurs may have a better idea about whether their business idea will actually work than the VC firm who must decide whether to provide the financial resources. This problem is exacerbated by the founder's incentive, being in possession of more information, to misrepresent it. *Ex ante* information asymmetry can lead to adverse selection risks, forcing VC firms to resort to detailed screening processes in order to distinguish between good-quality and poor-quality businesses. *Ex post* information asymmetry arises because an entrepreneur, having obtained the money, may not have sufficient commitment to the project, or may even use the funds in a way contrary to the initial agreement with the VC investor. This moral hazard problem forces VCs into considerable efforts to monitor their investments, using restrictive covenants and financial contracts, taking a seat on the investee firm's board of directors, as well as other related practices.

Despite wide recognition that VC firms have developed specialised abilities to select and monitor entrepreneurial projects, there are a number of reasons why implementing ESG standards in the VC investment process can provide further help in overcoming both adverse selection and moral hazard problems.

It is worth noting at this point that the rationale behind the association between ESG and information asymmetry in the VC market relies primarily on reputation-building theory (Cui et al. 2018), which asserts that corporate social responsibility (CSR) firms are characterised by a better information environment. So, to the extent that being ESG-compliant leads them to select and monitor firms that are more transparent and less opportunistic, VC firms would thereby suffer fewer problems of information asymmetry. In this regard, the existing literature recommends that, in order to exploit the reputational benefits associated with CSR engagement, firms should create a healthy information environment (Albinger and Freeman 2000; Greening and Turban 2000). Consistent with this, other studies point out that managers of firms with high CSR engagement exploit CSR disclosure as a tool to advance information transparency (Dhaliwal et al. 2011). For example, Clarkson et al. (2008) find a positive association between CSR performance and the overall level of discretionary disclosures in environmental and social reports or related web disclosures. This evidence is consistent with voluntary disclosure theory (Verrecchia 1983), which suggests that firms have an incentive to disclose high performance in order to differentiate themselves from competitors with poor performance; this, in turn, addresses potential adverse selection problems. Previous studies (Sethi et al. 2016) also support the reputation-building theory, documenting that managers of high CSR firms are more likely to improve the quantity and quality of information disclosure, which also applies to the quality of financial reporting.

CSR disclosure is a valuable type of non-financial disclosure giving investors relevant information. A typical CSR report covers "a great deal of information, such as expenditures related to environmental protection and climate change, charity donations, and employee welfare – all of which are typically not reported in financial statements but bear significant implications for assessing firm value" (Dhaliwal et al. 2014, p. 2). Because CSR disclosure provides third parties with valuable non-financial information, it is no surprise to find direct evidence (Egginton and McBrayer 2019) of a reduction in information asymmetries for public companies with transparent CSR disclosure strategies. This evidence is based on the authors' documentation of improvements in equity market liquidity.

On the subject of managers controlling the quality of information, Kim et al. (2012) examined whether firms engaging in CSR are more inclined to convey more transparent and reliable financial information to investors compared to other firms. They find that CSR firms are less likely to manage earnings through discretionary accruals, less likely to manipulate real operating activities, and less likely to be the subject of investigations by financial market authorities. Drawing on this evidence, the authors argue that ethical concerns are likely to drive managers to produce high-quality financial reports.

Kim et al.'s (2012) view is supported by Lopatta et al. (2016), who suggest that CSR activities can improve a firm's credibility by raising its moral standards, its quality of financial reporting, and its transparency. This should therefore lead to lower information asymmetry because higher trustworthiness is correlated with a lesser amount of hidden information and hidden characteristics. In other words, if managers are less opportunistic and desist from taking advantage of private information, their behaviour implies a high degree of trustworthiness which reduces the amount of hidden information and lowers monitoring costs.

In general, the existing literature assigns to CSR engagement the important role of reducing information asymmetries between a firm's internal and external actors. One important extension of the reputation-building hypothesis is that firms that face high reputational risks are subject to severe adverse selection problems and incur high monitoring costs, which is the case for those in which VCs typically invest, and this provides a strong incentive for these companies to improve the quality and quantity of the information they offer to third parties. This point is of special relevance here, because it leads to the conclusion that adopting an ESG policy should lead to greater information-related benefits for VC-backed firms compared to non-VC-backed firms.

5.4.3 *Are ESG-Compliant VC Firms More Likely to Add Value to Investee Firms?*

While the existing literature provides indirect evidence that implementation of ESG standards by VC firms may improve their screening and monitoring activities on account of the mitigation of *ex ante* and *ex post* information asymmetry, the question remains open of whether it constitutes a lever for adding value to investee companies. The basis for postulating such an association is the possibility that VCs who encourage portfolio companies to improve their ESG engagement might in turn see an increase in their own firm's value. However, previous studies on the relationship between a firm's ESG engagement and its financial performance have delivered mixed results and led to conflicting theoretical frameworks.

Existing empirical studies differ both with regard to the financial performance indicators used as a dependent variable—which vary from market measures (namely, Tobin's Q) to accounting-based indicators (e.g. ROA [return on assets] and sales growth)—and with regard to the reported findings. Some studies report a negative relationship (Bowman and Haire, 1975) a) or a non-significant one (McWilliams and Siegel 2001) between CSR engagement and financial performance. Many papers conclude a positive and linear relationship (Orlitzky et al. 2003), while others show a curvilinear effect of ESG engagement on firm financial performance (Barnett and Salomon 2012). Overall, findings thus far are not conclusive and leave the fundamental question still open: is ESG engagement beneficial or detrimental to a firm's enterprise value? As concluded by more recent research (Buchanan et al. 2018), the main challenge for academics is to address the endogenous nature of firm financial performance and CSR activities. However, studies that control for endogeneity also deliver mixed results. Emblematic of these is the research conducted by Awaysheh et al. (2020), who examined the relationship between ESG performance and financial performance by benchmarking firms against industry peers to identify best-in-class and worst-in-class. They find that the best-in-class firms outperform their industry peers in terms of operating performance and have higher relative market valuations. However, when they control for endogeneity, they continue to find the best-in-class firms getting higher market valuations than their industry peers, even though the significant relationship between operating performance and CSR categories disappears.

From a theoretical perspective, the main interpretative frameworks rely on the agency model (Friedman 1970), stakeholder theory (Russo and Perrini 2010) and the resource-based view (Barney 1991). The presence of different frameworks speaks not only to different ideas about how a firm functions but also to the lack of convergence of empirical evidence on causality between ESG and financial performance (Gangi et al. 2020). From an agency perspective, firms that channel some of their (financial and non-financial) resources towards ESG initiatives are less competitive than their competitors who do not divert any resources away from the pursuit of the exclusive goal of shareholder value maximisation. In fact, for academics who support the agency view, having numerous purposes beyond profit optimisation is equivalent to having no goals at all (Jensen 2001). By logical extension, it becomes more difficult to evaluate managers' performance, due to the high levels of discretion they exercise, which in turn can lead to distortions in their investment choices. Therefore, if we allow the possibility that managers over-invest in ESG initiatives in pursuit of their own self-interest rather than corporate goals, this supports the hypothesis intrinsic to the agency model that high ESG firms are more likely to suffer from loss of value. According to the over-investment hypothesis, a manager may also engage in CSR activities to cover up the impact of corporate misconduct (Hemingway and Maclagan 2004).

Although the above-mentioned studies focus on negative externalities deriving from ESG engagement, the debate has also been fed by alternative interpretative models that concern themselves with the potential benefits of adopting an ESG policy. ESG initiatives have progressively gained strategic relevance, which is in alignment with stakeholder theory and the view of a firm as a set of unique resources (Barney 1991). Subsequent advances in ESG literature have suggested a remodelling of the meaning of ESG engagement from a non-market activity into a social-market activity. From this perspective, good management implies the improvement of relationships with the various stakeholder groups: the better a firm manages these multiple relationships, the more competitive it will be in the long run (Waddock and Graves 1997) thanks to the accruing of social capital and trusting networks which minimise the costs of contracting (Barnett and Salomon 2012).

Regardless of whether engaging in ESG initiatives allows firms to establish closer relationships with their stakeholders, it has been shown to improve firms' chances of maximising their enterprise value. Some studies point to the higher productivity of employees (Asrar-ul-Haq et al. 2017) and the recruitment and retention of better-qualified human resources (Hartenian and Gudmundson 2000), while others point to higher efficiency deriving from environmental protection policies (Bringer and Benforado 1994) and improved access to financial resources (Goss and Roberts 2011). Still others point to the benefits arising from stronger customer loyalty, positive firm image and incentives for innovative expansion of business model activities (Yoon et al. 2006; Dangelico and Pujari 2010; Deng and Xu 2017). From a resource-based view, outcomes that derive from ESG actions may be intangible and firm-specific assets that link the investment in ESG to better financial performance (Gangi et al. 2020).

Overall, drawing on prior literature, we can conclude that not only does the adoption of an ESG investment policy represent no obstacle to the function of value creation traditionally played by VCs, but it may in fact give VC firms extra opportunities to maximise their portfolio firms' enterprise value.

5.4.4 ESG Considerations at Each Stage of the Investment Process: Some Operational Indications

In the previous section, we described the potential benefits deriving from the implementation of ESG criteria in each phase of the VC fund investment process. In this section, we provide insights about VC firms' practices that imply a high commitment to ESG standards. It should be understood that the analysis that follows of ESG for each individual investment phase complements, rather than replaces, VCs' traditional investment activities.

- **Screening.** A VC firm's investment process starts with an initial screening carried out on a large group of potential investee firms. It is in search of a small number of companies that can be subjected to a more accurate assessment. At this stage, the GP's investment team must first evaluate whether the investee falls foul of their exclusion list: if it belongs to a controversial sector or takes part in activities excluded by the VC fund's ESG policy, the investment cannot proceed, unless the team believes that the firm can become compliant with its ESG policy over a reasonable time frame. During this initial screening, it is also critical that the team assigns potential target firms with an ESG risk rating, in order to establish the degree of attention each one needs during due diligence.

- **Due diligence.** Whereas the screening stage could be described as just an initial skimming of potential investments, the due diligence process is where the investment team carefully gauges the firm's compliance with the VC fund's ESG policy. Due diligence normally includes, first, an analysis of publicly available information with the aim of controlling for business integrity and uncovering any environmental and social controversies. Second is a review of relevant information available on its corporate websites, corporate reports and other public company sources. A third customary process is to interview a firm's top management, shareholders, board members, employees and other main stakeholders. If the investment exposes the VC fund to high risks from an environmental or social perspective, the investment team normally requires an ESG assessment to be conducted by an external advisor. At this stage, it is also important to assess the quality of the company's management systems in relation to any ESG risks identified and propose improvements. If due diligence reveals poor ESG standards, the VC fund's investment team in collaboration with the management of the company under scrutiny, ultimately supported by the ESG advisor, need to identify appropriate remedial action. The due diligence ends with the preparation of the investment paper that enumerates all significant ESG issues and lays out an ESG action plan with a focus on costs and time frame.
- **Investment decision and negotiations.** A VC firm's investment decision is made by the investment committee generally based on the results of the ESG analysis (as reported in the investment paper) combined with those from the other due diligence areas (tax, legal and accounting, among others). If the decision is positive, then the investment negotiations begin, during which the action plan for improvements is discussed and shared by the VC fund's investment team and the management of the firm under scrutiny. An important part of the investment negotiations is the timely establishment of whether the investee firm has an ESG policy in line with its own

guidelines. In any case—and always before the investment is made—the fund manager should require a formal legal commitment by the target firm to comply with the VC fund's ESG policy.

- **Holding period.** During the holding period, the VC firm monitors its investment from an ESG perspective by checking, first, that the target firm is adhering to the VC fund's ESG policy and ensuring the implementation of any action plan for ESG improvements. Second, VC investors should ensure that newly identified risks and improvement opportunities are being properly addressed by portfolio companies; they should also make a record of any serious environmental and social incidents involving these companies. VC firms should monitor the main ESG performance indicators and disclose them to their LPs, and the general public, via their website and annual reports. Finally, they should also encourage portfolio firms to revise their strategies in response to changes in regulations, markets and technology that might occur during the holding period.
- **Exit.** The VC fund must take an interest in whether the ESG standards and practices implemented over the holding period will persist after its exit. Otherwise, the institutional investors cannot accomplish the dual goal of exploiting ESG efforts to preserve and increase their reputation and also produce good results for society as a whole. This is especially true for exits accomplished via IPO, which are thoroughly scrutinised by market participants both before the IPO and after.

5.5 When a VC Becomes a Social VC

In the previous sections, we discussed how conventional VC funds work, how they are beginning to incorporate ESG standards, and the potential effects associated with the inclusion of ESG guidelines within their investment process. In this section, our gaze turns towards social VC investments,[8] which have things in common with conventional VC

[8] The term "social VC" is often used as a synonym for "impact VC", while many academics and practitioners consider green VC to be a subset of social VC. For example, Randjelovic et al. (2003) define green VC as "financial capital provision invested in high-risk environment-oriented ventures, which offers the possibility of ecologically sound business practices, as well as significant gains to compensate for the risks involved in such investments" (p. 251).

investments as well as differences. In common to both, the capital injection is accompanied by the institutional investor's active involvement in the strategic management of the portfolio companies. Social VC therefore differs from traditional philanthropic initiatives because it provides investee firms with a range of non-financial services that includes strategic, marketing and financial advice and access to a wide network of professionals, financial intermediaries and entrepreneurs operating in the same area (Perrini and Vurro 2010). Where social VCs depart from traditional VCs is that, rather than having optimisation of the financial risk–return relationship as the key goal, their primary purpose is to maximise the social impact of their investments. Social VC firms, therefore, seek companies addressing environmental and social challenges across the globe, such as poverty, climate change, education, health and well-being, or financial inclusion. Companies financed by social VC funds might be tackling such pressing issues either directly through their products or services or by implementing specific programmes in these areas. An interesting example of social VC investment is that of Acumen, one of the world's most famous social VC funds, in Acceso Colombia, an agriculture aggregator and distributor linking Colombia's smallholder farmers with markets. Acceso Colombia buys high-quality fruit, vegetables and fish from farmers, processes and packages the produce in community-run warehouses, and delivers to high-value retailers and supermarkets. By removing intermediaries from the value chain, Acceso Columbia has been able to increase farmers' income from their yields.

Of course, it makes little sense for a social VC fund to create social impact if it is not measurable and communicable to those who provide them with financial resources. In fact, a core issue for social VC funds in general is the measurement and management of social and environmental outcomes alongside financial returns. To this end, both international organisations and private entities have been making considerable inroads into developing shared practice on impact evaluation and management. Two outcomes of this are the OECD Social Impact Investment Initiative and the Global Impact Investing Network (GIIN)'s Navigating Impact project. In parallel with the development of common frameworks for impact assessment, a growing number of social impact measurement tools have been proposed by academics and practitioners,[9] a prominent

[9] For further information, see Maas and Liket (2011).

one being "social return on investment" (SROI). Introduced by the US Roberts Enterprise Development Fund, SROI focuses on the principles of accountancy and cost–benefit analysis that assign monetary values to social and environmental returns to demonstrate wider value creation (Rotheroe and Richards 2007). This measures the value of social benefits created by an organisation in relation to the relative cost of achieving those benefits (see Eq. 5.1). The result is a ratio of monetised social value.

$$\frac{SROI = NetPresentValueofBenefits}{NetPresentValueofinvestment} \tag{5.1}$$

Although a number of successful examples of SROI can be cited (Ryan and Lyne 2008), in practice only a few managers and institutional investors, including VC funds, use it because of practical and implementation problems (Millar and Hall 2013). For example, an assessment of the net present value of benefits can be skewed by considerable estimation errors because it requires a counterfactual analysis of what might have happened without the social VC investment.

Even though social VC funds support socially responsible companies, they are far from being mere benefactors. In fact, like traditional VC funds, they select investment opportunities with a target financial return in mind. Like other impact investors, social VC funds' financial return expectations comprise a wide spectrum. According to the classification adopted by both JP Morgan/GIIN Impact Investor and Wharton Social Impact Initiative (WSII), it is possible to distinguish between social VC funds that seek market-rate returns, those that seek returns below but close to market rate, and those that seek returns closer to capital protection. Interestingly, the most recent JP Morgan/GIIN Annual Impact Investor survey (2020) finds that 81% of PE-focused investors—which includes social VCs—are seeking market-rate returns.

In the light of this evidence, it comes as no surprise to see the impact investing literature investigating whether social VC investors are penalising financial returns in order to create impact. However, no clear consensus on this issue emerges. Kovner and Lerner (2015) analysed the performance of a sample of social VC funds and found that their investments are less likely to succeed than investments conducted by traditional VC funds. Barber et al. (2021) benchmarked with analogous VC funds and controlled for a number of fund-specific characteristics to find that social VC funds get internal rates of return that are approximately 5%

lower. However, the WSII found evidence that market-rate-seeking VC funds are competitive with the market on a gross basis, and argued that they might also be financially competitive on a gross basis with other equity investing funds.

Social VC funds vary with regard to the targeting of financial returns and measuring social impact, but they may also vary significantly according to investment strategy and operative model (Perrini and Vurro 2010). To define a social VC fund's strategic orientation, first the areas of intervention must be demarcated, mainly in terms of social sector, geographical focus and firm life cycle stage. To illustrate how social VC funds can differ in terms of strategic orientation, Table 5.2 lists the areas of intervention of some of the world's most famous social VC funds. As can be seen, social VC funds tend to be highly specialised in terms of social sector, with more generalist funds such as BonVenture very much in the minority.

The choice of a clear strategic orientation confers a number of advantages on a social investment fund. First, with the number of impact investing funds increasing every year, according to the GIIN's Annual Impact Investor Survey, it is all the more important for a social VC fund to differentiate itself via a compelling and clearly articulated investment strategy. As with traditional VC firms, differentiation is fundamental in consolidating its ability to raise funds in an increasingly crowded market. Second, a clear investment strategy is necessary to allow social VC funds and investors to share, a priori, investment models and objectives. This is critical in avoiding conflict among the various entities involved and in reducing the risks associated with the cultural distance that often distinguishes promoters of social VC funds (Wagner 2002; Perrini and Vurro 2010). Furthermore, it is also crucial to properly define the social VC fund's organisational structure and the financing instruments to be used and to identify the ideal profile of the target investee organisations.

Notwithstanding the above considerations, a social VC fund's investment strategy can be strongly influenced by the profile of its promoters. Among the most active promoters in the social VC panorama are undoubtedly foundations, which are accustomed to combining the launch of a social VC fund with their traditional philanthropic activities. For example, the social VC fund Inspiring Scotland was launched in Scotland by the Lloyds TSB Foundation, although the former differs from the latter in its strategic orientation. A private firm might launch a VC fund simply to announce the strategic nature of its social commitment, or

Table 5.2 Investment focus of some world-famous social VC funds

Name and brief mission statement	*Geographical focus*	*Sector focus*
Acumen is a nonprofit global VC fund that invests in ambitious entrepreneurs and early-stage innovators tackling poverty.	East Africa, West Africa, India, Pakistan, America	Agriculture, education, energy, financial inclusion, health, housing, water, workforce development
Adobe Capital is a social fund focused on supporting the early growth of social and environmental small and growing businesses with a specific emphasis on companies operating in underprivileged base-of-the-pyramid communities.	Latin America	Agribusiness, base of the pyramid, water, recycling, clean tech, energy, ecotourism
BonVenture is the first social venture capital fund focusing on social enterprises in the German-speaking area.	German-speaking countries	No focus
The **European Fund for Southeast Europe (EFSE)** aims to foster economic development and prosperity in southeast Europe and in the European Eastern Neighbourhood region through the sustainable provision of additional development finance, notably to micro and small enterprises and private households, via qualified financial institutions.	16 emerging European countries	Microfinance institutions

(continued)

Table 5.2 (continued)

Name and brief mission statement	*Geographical focus*	*Sector focus*
Root Capital is a nonprofit social investment fund that looks for those places where environmental vulnerability and unmet financial need are greatest. Root Capital finds businesses that are doing the most to address these conditions and supports them to do even more.	Central America and Mexico, East Africa, South America, West Africa, Southeast Asia	Livelihood, women, environment, youth, peace, food security
The **Grassroots Business Fund (GBF)** strives to create a world where economic opportunity reaches everyone. GBF builds and supports high-impact businesses that provide sustainable economic opportunities to millions of people at the base of the economic pyramid.	Africa, Asia, Latin America	Agribusiness, handcrafted manufacturing, mobile banking, clean energy, lighting, technology, heating
The **Oasis Fund** invests in commercially viable companies that deliver essential goods and services directly benefiting low-income communities by providing access to affordable housing, healthcare, education, energy, livelihood opportunities, water, sanitation and the like.	Africa. Asia, Central America	Agriculture, housing, healthcare, education, energy, financial services

Name and brief mission statement	*Geographical focus*	*Sector focus*
The **Global Partnerships/Eleos Social Venture Fund (SVF)** is an impact-first fund investing in early-stage social enterprises to expand opportunities for people living in poverty in Eastern Africa. The fund was created to address the "pioneer gap" by supporting early-stage social enterprises with a combination of investment capital and board-level advisory support.	Eastern Africa	Education, energy, health, water, sanitation, livelihoods

Source Information retrieved from social VC funds' websites as a strategy to differentiate its business, which is sometimes the case for large multinational companies or traditional PE and VC firms. A prominent example of this is Impact Ventures UK which was launched in 2013 by Berenberg Investment Bank. Funds might also be launched as an initiative by an individual philanthropist placing part of his/her accumulated wealth at the service of a social cause, accompanying this move with a direct commitment to the fund created. This is the case with Bridges Ventures, a pioneering social investing firm in the UK, and the Italian Oltre Venture, each founded by an entrepreneur with long-term experience in the VC and PE industry. Finally, a promoter might be an agency or a public institution. As Perrini and Vurro (2010) point out, public bodies rarely become direct sole promoters of a social VC fund: they are more likely to intervene through a partnership or cooperative agreement with another for-profit or nonprofit organisation. A recent example is the India Inclusive Innovation Fund, launched in 2014 by the National Innovation Council, in partnership with the Ministry of Micro, Small and Medium Enterprises, with the aim of investing in ventures catering to the country's poor.

Social VC funds are highly heterogeneous with regard not only to their strategies and their promoters but also their operating models. Specifically, they are differentiated by the various financial instruments and financial and non-financial services they offer to their funded companies (Perrini and Vurro 2010). As highlighted by Achleitner et al. (2011), the range of funding options includes the following types of instruments:

- **Grants**, an important funding option in the social sector, are typically provided by foundations or individuals. Although this type of financial aid does not have to be repaid, grants are not without their drawbacks. First, they are awarded only for certain projects and usually exclude overhead costs and expenditures. Second, they are usually short-term, hard to predict, and impose high fundraising costs on the firm.
- **Debt** instruments can be used when a social business can be expected to generate stable and predictable cash flows, so it can pay the annual interest as well as eventually repaying the capital (generally after between five and seven years).
- **Equity** is the financing instrument posing the highest risk for social VC investors. Here, they provide the investee firm with a certain amount of finance in exchange for an equity stake, which, as is well

understood, allocates it a portion of the dividends that the social VC-backed firm will distribute, but no regular annual payment. The investor will also have some control and voting rights, which are usually defined in the contract between investor and investee.

- **Mezzanine capital** combines elements of debt and equity instruments, and is an appropriate funding option when neither of these alone is suited the characteristics of the investee company. The interest payment can be linked to the firm's earnings, while the reimbursement takes place after an agreed number of years with the repayment of the capital or its conversion into an equity stake. Its useful elasticity makes mezzanine capital an appealing funding option for both social VC investors and social firms.
- **Hybrid capital** combines the characteristics of grants, debt and equity. Of key importance here is that hybrid capital can be converted into a grant in certain well-defined scenarios. Social VC funds use a variety of financing instruments that can be characterised as hybrid capital. A *recoverable grant* is a loan that is repaid when the firm achieves pre-agreed goals; if the goals are missed, it is converted into a grant. This instrument can be used when it is predicted that the project's success would allow the firm to repay the loan. A *forgivable loan* is based on a converse logic: here, if the firm attains its pre-agreed milestones, the loan does *not* have to be repaid and is converted into a grant. A *convertible grant* is one that will be transformed into an equity stake in the case of success. Finally, *revenue share agreements* are royalty-based loans that allow the investor to finance a firm in exchange for a percentage of future revenues; this risk-sharing model provides the social firm with high financial flexibility.

5.6 The Dark Side of VC: Addressing the VC Gender Gap

Attracting VC funding has always been a major challenge for female entrepreneurs (Coleman and Robb 2012; Balachandra 2020). The DIANA Project report was a pivotal analysis of women's access to VC funds in the US, revealing that investment in firms with a female on the team had never exceeded 4% of total value invested over the period 1988–1999 (Brush et al. 2018). It also found that these few recipients were largely concentrated in certain sectors (biotechnology, medical/health and software) and were more likely to attract early-stage financing.

Subsequent studies have documented some progress in recent years, although the VC industry gender gap remains.[10] For example, Brush et al. (2018) report that the percentage of companies attracting VC funding that have women on their executive team almost quadrupled during the early 2000s—from fewer than 4% of total VC investments in 1999 to 15% over the period 2011–2013. Data sources such as Crunchbase have recently reported that, in the US, only 3% of VC funds find their way to all-female teams, with 8% going to mixed teams and the remaining 89% to all-male teams (Crunchbase News 2019). In the VC industry, the lack of female representation is evident not only among investee companies but also inside VC firms themselves, as highlighted by Balachandra (2020), who finds that the 93% of US VCs are men. Likewise, the recently published *New York Times* and CB Insights 2018 list reveals that only nine out of the 100 top VC partners worldwide are women.

The persistence of this gender gap in the VC industry—which, furthermore, runs counter to more general labour market trends (Goldin 2006)—has triggered investigations into the main drivers of this phenomenon. Brusher et al.'s (2018) is a thorough theoretical framework which offers three main theoretical explanations for the disparity. The first relies on the idea that the gender gap arises from the nature of social networks within the VC industry. Previous studies show that both VC firms and those who come to them with deals (angel groups, research labs, consulting firms, universities, financial intermediaries) are strongly networked (Sorenson and Stuart 2001). These connections, which are made even tighter by the small number of VCs and their concentration in certain geographical areas (e.g. Silicon Valley), mean that information about VC investment opportunities is passed around within a few elite circles. As a result, the possibility of entrepreneurs being closely connected to these networks influences their likelihood of receiving VC funding. Female entrepreneurs are therefore penalised, because of the male homogeneity of the VC industry and the likelihood that they will have different networks from their male counterparts).

[10] For this reason, the National Venture Capital Association (NVCA) (2014), the largest VC industry organisation in the US, formed a task force to promote greater inclusion of women in both VC firms and investee companies.

The second explanation looks to the unique characteristics of the VC industry's institutional environment. The complex integrated web of rules, beliefs and practices developed in the VC industry to minimise information asymmetries between principals and agents are predominately male gendered (Becker-Blease and Sohl 2007). For example, the rules for pitching and coaching tend to be confrontational, competitive and judgemental, characteristics that favour established ideas and norms of masculinity (Eagly and Karau 2002).

The third explanation is that VC firms, who are predominantly male, are influenced by stereotypes and gender homophily. A VC's decision on whether to invest in a firm will to some extent depend on the entrepreneur's pitch, during which questions will typically be asked about the problem being addressed, unique advantages of the business concept, and the market size and growth, whereupon the VC will form a perception of the initiative's likely success (Baron 1998). Various studies (*inter alia*, Eddleston and Powell 2008) suggest that, in an interview situation such as this, value judgements are inextricably linked to observable factors (e.g. stereotypical perceptions about character and integrity) which are typically associated with male traits. Female entrepreneurs therefore suffer by not conforming to these stereotypes during their pitch. In addition to this, homophily is also at play, which is the tendency for individuals to associate—i.e. perceive positively and trust—those who are demographically similar, and this applies to gender (Brashears 2008). Applied to the male-dominated VC industry, with its severe information asymmetry, the gender homophily theory offers a further explanation of why VC firms prefer companies with male CEOs or all-male teams.

Moving on from the theoretical premises, we see that, from an empirical point of view, the literature is heading in various directions. Guzman and Kacperczyk (2019) indicate that gender disparity arises at different stages of the entrepreneurship process, with the most noticeable stage being the initial startup: women are less likely to found firms that signal growth potential to external investors. The blame for the remainder of the gender gap is placed on VC firms' biases and gender stereotypes. Ewens and Townsend (2020) find that female entrepreneurs are significantly less successful in attracting interest from male VC investors compared to observably similar male entrepreneurs. In contrast, those same female entrepreneurs are more successful than their male counterparts when they confront female investors. As these findings are not skewed by differences in entrepreneur gender in the main characteristics of startups, the authors argue that male and female VC investors are symmetrically biased in favour of their own gender, presumably with gender homophily coming into play here. Departing from previous studies, Alnamlah (2020) rejects the hypothesis that gender disparity in VC funding is driven by discrimination. She undertook a randomised field experiment to observe the behaviour of VC investors towards female entrepreneurs during the screening stage of funding proposals. Her finding is that male VC investors show women no differential treatment, and yet female entrepreneurs are at an advantage when evaluated by female VC investors.

Despite the multiple factors identified by all these studies, they converge on the idea that increasing the number of female VC investors will lead to a bigger share of VC funds making their way towards female businesses and will help to close the VC gender gap. Drawing on this conclusion, the literature has started to investigate whether an increase in the number of female investors might lead to a change in VC industry performance, with repercussions both for the economy and for LPs' revenue. Gompers et al.'s (2014) innovative study explored gender in relation to the performance of a sample of US VC investments to find that female VCs exhibit investment performance that is around 10–15% lower than their male colleagues. However, they provide evidence that this performance gap is not attributable to relative skills but mainly down to the fact that female VC investors accrue less benefit than male VC investors from their co-investment partners' track records.

Overall, the existing literature suggests that overcoming the combination of men's gender bias and the lack of female representation among VC investors and founders is a major challenge for the VC industry in the coming years.

References

Achleitner AK, Heinecke A, Noble A, Schöning M, Spiess-Knafl W (2011) Social investment manual: an introduction for social entrepreneurs. Available at SSRN 1884338

Albinger HS, Freeman SJ (2000) Corporate social performance and attractiveness as an employer to different job seeking populations. J Bus Ethics 28(3):243–253

Alnamlah MS (2020) Is the gender gap in venture funding driven by biased investors? In Academy of management proceedings (Vol. 2020, No. 1, p. 20763). Briarcliff Manor, NY 10510: Academy of Management

Asrar-ul-Haq M, Kuchinke KP, Iqbal A (2017) The relationship between corporate social responsibility, job satisfaction, and organizational commitment: Case of Pakistani higher education. J Clean Prod 142:2352–2363

Awaysheh A, Heron RA, Perry T, Wilson JI (2020) On the relation between corporate social responsibility and financial performance. Strateg Manag J 41(6):965–987

Balachandra L (2020) How gender biases drive venture capital decision-making: exploring the gender funding gap. Gend Manag

Balboa M, Marti J (2007) Factors that determine the reputation of private equity managers in developing markets. J Bus Ventur 22:453–480

Baldini M, Dal Maso L, Liberatore G, Mazzi F, Terzani S (2018) Role of country-and firm-level determinants in environmental, social, and governance disclosure. J Bus Ethics 150(1):79–98

Barber BM, Morse A, Yasuda A (2021) Impact investing. J Finance Eco 139(1):162–185

Barnett ML, Salomon RM (2012) Does it pay to be really good? Addressing the shape of the relationship between social and financial performance. Strateg Manag J 33(11):1304–1320

Barney J (1991) Firm resources and sustained competitive advantage. J Manage 17(1):99–120

Baron RA (1998) Cognitive mechanisms in entrepreneurship: why and when enterpreneurs think differently than other people. J Bus Ventur 13(4):275–294

Becker-Blease JR, Sohl JE (2007) Do women-owned businesses have equal access to angel capital? J Bus Ventur 22(4):503–521

Botsari A, Lang F (2020) ESG considerations in venture capital and business angel investment decisions: evidence from two pan-European surveys (No. 2020/63). EIF Working Paper

Bowman EH, Haire M (1975) A strategic posture toward corporate social responsibility. Calif Manage Rev 18(2):49–58

Brashears ME (2008) Gender and homophily: differences in male and female association in Blau space. Soc Sci Res 37(2):400–415

Bringer RP, Benforado DM (1994) Pollution prevention and total quality environmental management. Environmental strategies handbook, 165–188

Brush C, Greene P, Balachandra L, Davis A (2018) The gender gap in venture capital-progress, problems, and perspectives. Ventur Cap 20(2):115–136

Buchanan B, Cao CX, Chen C (2018) Corporate social responsibility, firm value, and influential institutional ownership. J Corp Finance 52:73–95

Bürer MJ, Wüstenhagen R (2008) Cleantech venture investors and energy policy risk: an exploratory analysis of regulatory risk management strategies. Sustainable innovation and entrepreneurship. Edward Elgar Publishing, 290–309

Campbell K, Mínguez-Vera A (2008) Gender diversity in the boardroom and firm financial performance. J Bus Ethics 83(3):435–451

Chemmanur TJ, Krishnan K, Nandy DK (2011) How does venture capital financing improve efficiency in private firms? A look beneath the surface. Rev Financ Stud 24(12):4037–4090

Clarkson PM, Li Y, Richardson GD, Vasvari FP (2008) Revisiting the relation between environmental performance and environmental disclosure: an empirical analysis. Account Organ Soc 33(4–5):303–327

Coleman S, Robb A (2012) Gender-based firm performance differences in the United States: examining the roles of financial capital and motivations. In Global women's entrepreneurship research. Edward Elgar Publishing

Croce A, Martí J, Murtinu S (2013) The impact of venture capital on the productivity growth of European entrepreneurial firms:'screening'or 'value added'effect? J Bus Ventur 28(4):489–510

Crunchbase News (2019) Inside VC firms: the gender divide, August 14, 2019

Cui J, Jo H, Na H (2018) Does corporate social responsibility affect information asymmetry? J Bus Ethics, 148(3):549–572

Cumming D, Johan S (2008) Information asymmetries, agency costs and venture capital exit outcomes. Ventur Cap 10(3):197–231

Cumming DJ, Johan SA (2013) Venture capital and private equity contracting: an international perspective. Academic Press

Dangelico RM, Pujari D (2010) Mainstreaming green product innovation: why and how companies integrate environmental sustainability. J Bus Ethics 95(3):471–486

Deng X, Xu Y (2017) Consumers' responses to corporate social responsibility initiatives: the mediating role of consumer–company identification. J Bus Ethics 142(3):515–526

Dhaliwal DS, Li OZ, Tsang A, Yang YG (2011) Voluntary nonfinancial disclosure and the cost of equity capital: the initiation of corporate social responsibility reporting. Account Rev 86(1):59–100

Dhaliwal D, Li OZ, Tsang A, Yang YG (2014) Corporate social responsibility disclosure and the cost of equity capital: the roles of stakeholder orientation and financial transparency. J Account Public Policy 33(4):328–355

Eagly AH, Karau SJ (2002) Role congruity theory of prejudice toward female leaders. Psychol Rev 109(3):573

Eddleston KA, Powell GN (2008) The role of gender identity in explaining sex differences in business owners' career satisfier preferences. J Bus Ventur 23(2):244–256

Egginton JF, McBrayer GA (2019) Does it pay to be forthcoming? evidence from CSR disclosure and equity market liquidity. Corp Soc Responsib Environ Manag 26(2):396–407

Ewens M, Townsend RR (2020) Are early stage investors biased against women? J Financ Econ 135(3):653–677

Friedman M (1970) The social responsibility of business is to increase its profits. The New York Times Magazine (September 13) (Retried on 13 January 2012 from: http:// www.colorado.edu/studentgroups/libertarians/issues/friedman-soc-resp-business. html)

Gangi F, Meles A, Monferrà S, Mustilli M (2020) Does corporate social responsibility help the survivorship of SMEs and large firms? Glob Finance J 43:100402

Goldin C (2006) The quiet revolution that transformed women's employment, education, and family. Am Econ Rev 96(2):1–21

Gompers PA, Lerner J (1999) What drives venture capital fundraising? (No. w6906). National bureau of economic research

Gompers P, Kovner A, Lerner J, Scharfstein, D (2008) Venture capital investment cycles: the impact of public markets. J Finance Eco 87(1):1–23

Gompers PA, Mukharlyamov V, Weisburst E, Xuan Y (2014) Gender gaps in venture capital performance. Available at SSRN 2445497

Goss A, Roberts GS (2011) The impact of corporate social responsibility on the cost of bank loans. J Bank Financ 35(7):1794–1810

Greening DW, Turban DB (2000) Corporate social performance as a competitive advantage in attracting a quality workforce. Bus Soc 39(3):254–280

Guzman J, Kacperczyk AO (2019) Gender gap in entrepreneurship. Res Policy 48(7):1666–1680

Hartenian LS, Gudmundson DE (2000) Cultural diversity in small business: implications for firm performance. J Dev Entrep 5(3):209

Hemingway CA, Maclagan PW (2004) Managers' personal values as drivers of corporate social responsibility. J Bus Ethics 50(1):33–44

Ioannou I, Serafeim G (2012) What drives corporate social performance? the role of national-level institutions. J Int Bus Stud 43(9):834–864

Jensen MC (2001) Value maximization, stakeholder theory, and the corporate objective function. J Appl Corp Finance 14(3):8–21

Kaplan S, Schoar A (2005) Private equity performance: returns, persistence and capital flows. J Finance 60:1791–1823

Kim Y, Park MS, Wier B (2012) Is earnings quality associated with corporate social responsibility? Account Rev 87(3):761–796

Kovner A, Lerner J (2015) Doing well by doing good? community development venture capital. J Econ Manag Strategy 24(3):643–663

KPMG Venture Pulse Q4 (2018) Global analysis of venture funding, January 15, 2019

Krishnan CNV, Ivanov VI, Masulis RW, Singh AK (2011). Venture capital reputation, post-IPO performance, and corporate governance. J Financial Quant Anal, 1295–1333

Lerner J, Tåg J (2013) Institutions and venture capital. Ind Corp Chang 22(1):153–182

Lopatta K, Buchholz F, Kaspereit T (2016) Asymmetric information and corporate social responsibility. Bus Soc 55(3):458–488

Maas K, Liket K (2011) Social impact measurement: classification of methods. In Environmental management accounting and supply chain management (pp. 171–202). Dordrecht: Springer

McWilliams A, Siegel D (2001) Profit maximizing corporate social responsibility. Acad Manage Rev 26(4):504–505

Megginson WL, Meles A, Sampagnaro G, Verdoliva V (2019) Financial distress risk in initial public offerings: how much do venture capitalists matter? J Corp Finance 59:10–30

Millar R, Hall K (2013) Social return on investment (SROI) and performance measurement: the opportunities and barriers for social enterprises in health and social care. Public Manag Rev 15(6):923–941

Orlitzky M, Schmidt FL, Rynes SL (2003) Corporate social and financial performance: a meta-analysis. Organ Stud 24(3):403–441

Perrini F, Vurro C (2010) Social venture capital & venture philanthropy. Modelli e processi d'investimento nell'innovazione sociale, Egea, Milano

Pitchbook (2020a), NVCA venture report, January 13, 2020

Pitchbook (2020b), European venture report, January 29, 2020

Pivato S, Misani N, Tencati A (2008) The impact of corporate social responsibility on consumer trust: the case of organic food. Bus Ethics 17(1):3–12

Porter ME, Kramer MR (2006) The link between competitive advantage and corporate social responsibility. Harv Bus Rev 84(12):78–92

PRI (2015) The limited partners' responsible investment due diligence questionnaire

Randjelovic J, O'Rourke AR, Orsato RJ (2003) The emergence of green venture capital. Bus Strategy Environ 12(4):240–253

Ryan PW, Lyne I (2008) Social enterprise and the measurement of social value: methodological issues with the calculation and application of the social return on investment. Education, Knowledge & Economy 2(3):223–237

Rotheroe N, Richards A (2007) Social return on investment and social enterprise: transparent accountability for sustainable development. Soc Enterp J

Ruiz-Jiménez JM, del Mar Fuentes-Fuentes M, Ruiz-Arroyo M (2016) Knowledge combination capability and innovation: the effects of gender diversity on top management teams in technology-based firms. J Bus Ethics 135(3):503–515

Russo A, Perrini F (2010) Investigating stakeholder theory and social capital: CSR in large firms and SMEs. J Bus Ethics 91(2):207–221
Sethi SP, Martell TF, Demir M (2016) Building corporate reputation through corporate social responsibility (CSR) reports: the case of extractive industries. Corp Reput Rev 19(3):219–243
Siegel DS, Vitaliano DF (2007) An empirical analysis of the strategic use of corporate social responsibility. J Econ Manag Strategy 16(3):773–792
Sorenson O, Stuart TE (2001) Syndication networks and the spatial distribution of venture capital investments. Am J Sociol 106(6):1546–1588
State Street Global Advisors (2019) Into the mainstream: ESG at the tipping point
Trester JJ (1998) Venture capital contracting under asymmetric information. J Bank Financ 22(6–8):675–699
Verrecchia RE (1983) Discretionary disclosure. J Account Econ 5:179–194
Waddock SA, Graves SB (1997) The corporate social performance–financial performance link. Strateg Manag J 18(4):303–319
Wagner L (2002) The 'new' donor: creation or evolution? Int J Nonprofit Volunt Sect Mark 7(4):343–352
Yoon Y, Gürhan-Canli Z, Schwarz N (2006) The effect of corporate social responsibility (CSR) activities on companies with bad reputations. J Consum Psychol 16(4):377–390

Conclusion

This book has been written to provide a thorough understanding of one of the most significant ongoing trends in the financial industry: namely, the growth of socially responsible investments (SRIs) and, more generally, the increasing attention that participants in financial systems are paying to environment, social and governance (ESG) issues.

As has become clear in the course of the foregoing discussions, the value of global sustainable investments has more than doubled in the last decade, and there is a broad consensus that it will continue to grow steadily because of the increasing interest shown not only by conventional funds but also by alternative investors such as private equity (PE) and venture capital (VC) funds. Furthermore, in addition to banks based on cooperative and ethical principles with close ties to local communities, the whole banking system has started to completely re-evaluate the sustainability issue in the light of its growing importance for various stakeholders. Emblematic of this is the European Central Bank (ECB)'s Banking Supervision commitment to make banks manage their climate-related risks more effectively and disclose them more transparently, underpinned by an awareness that climate change threatens the financial system's very stability. Finally, we have demonstrated that the increasing interest in ESG issues shown by institutional investors and credit intermediaries is being accompanied by an emerging take-up among retail investors, which, in turn, is favoured by the growing diffusion of ESG ratings and innovative and more inclusive ways of financing green and social projects. In this

F. Gangi et al., *The Evolution of Sustainable Investments and Finance*,
https://doi.org/10.1007/978-3-030-70350-9

scenario, the democratisation of financing through crowdfunding is an additional tool contributing to the engagement of stakeholders concerned about social issues.

These signs of radical change emanating from the financial industry have motivated us to centre this book primarily around how different financial institutions are adapting their business practices to deal with ESG issues. These signs of change have, in addition, thrown up an array of searching questions; in responding to these we have attempted to draw our own conclusions about the evolution of sustainable investments and finance.

First of all, why exactly is the financial industry focusing more on sustainable development goals and ESG issues? We have identified several factors, a prominent one being a growing collective awareness in capital markets that a financial industry focused only on short-term returns cannot achieve its mission of allocative efficiency and guarantee investors an optimised risk-adjusted long-run performance. The 2007–2008 global crisis, the effects of which still persist in a number of countries and sectors, has ushered in a new era in which a more sustainable long-term approach is increasingly being preferred over the frantic search for short-run returns. There has been considerable progress among the literature studying the economic viability of ESG investments, and this has further served to propagate the idea that the development paradigm, and finance, has to change. The traditional "doing well but not good" rubric, which has characterised the prevailing speculative finance model of recent decades, is giving way to a new vision in which there is no dichotomy between financial returns and the contribution of investment to sustainable development: "doing well by doing good". In this regard, our literature review here has indicated a positive relationship between social and financial performance and, furthermore, implies that superior risk-adjusted returns are available to investors who integrate traditional asset allocation indicators with ESG criteria. This is an idea that emerges clearly throughout the chapters in this book to the extent that, in each financial sector analysed, multiple reasons have been identified why being ESG-compliant can help financial intermediaries do good while doing well. The consistent growth in ESG investments has also been enabled by a growing number of initiatives from governments and regulators worldwide aimed at promoting a transition to a more sustainable society. A prominent example is the recent legislative package proposed by the European Commission to encourage sustainable investment within the

EU financial system. The COVID-19 recovery plan allocates around 33% (€356 billion) of the EU's long-term budget (2021–2027) to natural resources and the environment, thus providing additional stimuli to sustainable investments and funding.

These considerations lead on to the second point in our conclusions. This concerns the main opportunities and risks that result from the changes that financial institutions are making to their business model in order to deal with ESG issues. From an institutional investor perspective, we have identified the risk that the transition of SRIs from niche to mainstream might mean an integration of socially responsible issues into conventional investment approaches but at the cost of losing the true ethical identity of SRI. Furthermore, we have shown that the implementation of ESG guidelines implies the fulfilment of two levels of fiduciary duty. On the one hand, fund managers are required to include in the investment portfolio firms that present both favourable corporate financial performance (CFP) and corporate social performance (CSP), in order to satisfy both the financial and extra-financial expectations of individual investors. On the other hand, the managers of investee companies are required to engage in ESG initiatives in order to meet the investment screening criteria of socially responsible investment funds (SRFs). This dual fiduciary mandate can duplicate agency costs by increasing the exposure of the investors to the risk of the SRF's fiduciary mandate not being fulfilled.

Nonetheless, remarkable opportunities can await institutional investors that integrate ESG guidelines into the investment process. Based on reputation-building theory, which asserts that firms compliant with corporate social responsibility (CSR) are characterised by a superior information environment, we have argued that an implementation of ESG guidelines leads institutional investors to select and monitor firms that are more transparent and less opportunistic. These informational benefits may be particularly pronounced in the case of alternative funds as they invest in contexts in which information asymmetry is extreme. Furthermore, drawing on the prior literature on the relationship between CSP and CSR performance, we can also conclude that the adoption of an ESG investment policy can assist the value creation function traditionally played by active institutional investors, thereby offering extra opportunities to maximise their portfolio market value. From our critical viewpoint, efforts must be made by institutions and policymakers to guarantee the certification of SRI in order to avoid agency conflicts at the expense of savers and society as a whole. The main conclusion of our book with regard to this issue is that a dichotomy between socially responsible and conventional investors is not consistent with the widespread concern that sustainability is no longer an option but a necessity. To reach such a destination, we believe that what is required is a definitive cultural leap and clear rules of the game. The emerging ESG ratings and the EU Taxonomy on Sustainable Finance represent tools that are assisting travel in the desired direction. Favourable regulatory trends, risk mitigation and the signalling of goals are all additional motivations for socially responsible investments.

This increasing diffusion of SRI intercepts the social and financial function of a core agent such as a bank, delivering twofold benefits. Indeed, from the banking system's perspective, our view is that the integration of ESG principles into business practices can allow banks to reclaim their social role by becoming supporters of sustainable development. Even if the banking system has relatively little *direct* involvement in environmental and social issues compared to other industries, because of its function as a credit intermediary it can deliver a "multiplier effect", potentially benefiting other economic sectors. The universal application of ESG criteria as part of the assessment of creditworthiness can contribute to the necessary cultural leap we identified above. We have also argued that, interestingly, the reattainment by banks of their social function is not incompatible with their entrepreneurial nature and from the perspective of significant contribution to sustainable development of economic and social systems.

In fact, based on existing studies we have identified multiple channels through which a bank's commitment to ESG can produce non-conflicting social and economic benefits.

In this book we have identified the opportunities resulting from the alternative financial models that have rapidly taken off across the world as innovative ways of financing both businesses and social projects. Technological development has led to a dual transformation in CSR engagement, in which we have seen the creation of both new spaces and new forms of dialogue between organisations and stakeholders. For example, digital spaces (e.g. social media, sharing platforms) allow organisations to directly monitor information flows to and from stakeholders, with significantly higher time and cost efficiency compared to traditional channels. This allows the collection of real-time feedback from stakeholders about, for example, CSR strategies and reports.

The last question which we aim to answer conclusively is as follows: Is the increase in ESG investments just a temporary trend rather than an expression of structural changes in the financial markets? Certain factors refute this and substantiate our claim that ESG investments are making a definitive transition from niche to mainstream. First, the expansion of the information set introduced by regulatory requirements for non-financial reporting, along with regulatory innovation, demonstrates that the structural nature of sustainable finance is one in which non-financial performance assessment is integrated into investment strategies.

Second, the recent increased market uncertainty due to the COVID-19 pandemic that emerged in the first quarter of 2020, and due also to environmental and social protests being staged around the world, has forced fund managers to face the full range of risks to which their portfolios are exposed. Many have thereby come to understand that they had long underestimated the true financial impact of environmental risk. This is a lesson that is unlikely to be forgotten; it will consolidate the profound changes taking place in the asset allocation industry.

The final point concerns the role of millennials, who are expected to definitively change the investment landscape on account of their preference for investing in alignment with their personal values; this demographic is destined in the coming years to receive several trillion dollars of inherited wealth. As a result, fund managers will have to allocate more resources to develop products that can capture this emerging client segment, and this change in financial product offering will help ensure that in future sustainable investments are no longer a niche but a central platform for financial and real investments.

The evolution towards a new paradigm as described in this book calls for renewed efforts among academics and practitioners. This is a generation with the extraordinary and complex responsibility that comes with being the women and men involved in training activities, companies' missions and financial institutions in this pivotal time in our history.

Index

A

add value to investee firms, 214, 218, 221, 222, 227, 229, 231, 232, 234, 240
advantages on a social investment fund, 236
agency theory, 2, 8, 9, 24, 25
alternative investment choices, 70
alternative investments, 59–64, 67, 70
American SRI market, 42, 44
AUM growth of SRI, 37, 38, 44

B

bank as source of innovation, 102
bank-centred economic systems, 105
bank crowdfunding, 186, 187
banking business peculiarities, 118, 138
banking interventions, 104, 105, 107
bank lending and CSR, 125
banks and ESG criteria, 110
banks and sustainable development, 111
banks business model, 109, 115, 139, 141, 143
banks clients, 113
banks economic functions, 104
banks human capital, 116
bank social disclosure (BSD), 128
bank social performance (BSP), 108, 117, 129
bank's reliability, 113
banks reputation, 113, 114, 125, 129
banks trust recovery, 114
barriers to finance, 168
benefits of ESG policy, 221, 223, 224, 229, 230, 233
benefits of sustainable real estate, 89
better risk–return combination, 23
brand recognition effect, 116
BSD and stakeholder categories, 128, 135
business diversification, 187, 188
business startups, 72

C

capital acquisition strategy, 141

F. Gangi et al., *The Evolution of Sustainable Investments and Finance*, https://doi.org/10.1007/978-3-030-70350-9

capitalist individualism, 6
changes in banks' economic role, 104
channels to liquidate VC stakes, 217
clients, 109, 111, 114–116, 133, 135, 140
climate change, 16, 35, 39, 40
climate change risk management, 90
climate damage, 4
co-creation experiences, 173–175
co-financing, 164, 188, 190
collaborative opportunities, 173
collective social capital, 172, 190
communitarian logic, 195
community development and investment, 92
community related initiatives, 127
community relations, 133
companies' business models, 80
company's organisational performance, 70
company's social policies, 84
concerns of ESG metrics, 31
construction-sector companies, 90
contract imperfections, 24
co-production (CP), 173, 174
corporate crowdfunding, 184
corporate engagement programme, 68
corporate responsibility, 7, 39
corporate social responsibility (CSR), 172, 176–179, 181–184, 190, 191, 194
creation of sustainable real estate, 92
crowd-based sustainable finance, 194, 196
crowdfunders motivations, 159, 169, 175
crowdfunding (CF), 156–196
crowdfunding models, 159, 167–170, 187, 191–194
crowdfunding platforms (CFP), 157, 158, 163, 165, 169, 175, 178, 179, 181, 183, 186–190
crowdfunding stages, 164
crowdfunding stakeholders, 175, 179
crowdinvesting, 160, 169, 170, 178, 186, 191
crowdlenders, 168, 180–182, 195
crowdsourcing, 156, 157, 171
CSR activities, 228–230
CSR and access to finance, 124
CSR and bank performance, 125
CSR and cost of debt, 124
CSR and financial performance, 112, 124, 126, 127
CSR dimensions and financial performance, 127, 128
CSR dissemination, 109
CSR engagement, 177, 178, 181, 183, 184, 195, 196, 227–229
CSR oriented crowdfunding, 184
CSR over-compliance, 112
customers and CSR banking, 113

D

deadline effects, 164
democratisation of finance, 157, 166, 179
deregulation, 4
developing sustainable businesses, 211
differences in entrepreneur gender, 244
diffusion of responsibility, 163, 164
digital CSR, 190
disinvestment channels, 74
divestment period and redistribution, 217
donation-based crowdfunding, 159, 194
dual level of information and monitoring, 26
due diligence analysis, 84
due diligence phase, 82, 83

E

economic crises, 5
economic development, 101–110, 113, 116–118
economic growth, 103, 107, 115
efficiencracy, 179
employees, 30, 69, 84, 115, 120, 130, 131, 133, 135–137, 178, 184, 192, 194, 231, 232
employment quality, 120
energy and water efficiency improvement, 88
engagement, 13, 15, 16, 18, 19, 27, 30, 33, 36, 38, 40, 43, 46
engagement strategy decline, 46
entities involved in a VC fund, 215
entrepreneurs, 157–159, 161–163, 165, 166, 168–170, 172, 175, 181, 182, 186, 188, 191–195
entrepreneur's purchasing power, 102, 104
environmental and social KPIs, 130
environmental and sustainable real estate investments, 87
environmental objectives, 35
environmental pillar, 131
environmental specific programs, 234
Equator Principles (EPs), 111–113, 130
equity crowdfudning, 160, 168–170, 172, 186, 189, 192, 193, 195
ESE and CSR principles, 87
ESG agencies, 30–32, 36
ESG aspects, 63, 75–79, 82, 83, 86, 91, 92, 225
ESG aspects in hedge fund investments, 85, 86
ESG asset management, 76
ESG commitments, 66
ESG compliance, 220, 223, 224, 232
ESG criteria, 59, 60, 65, 70–72, 77, 78, 80, 81, 83, 84, 86, 92
ESG due diligence, 60, 75, 78, 81, 82
ESG engagement, 220–223, 229, 230
ESG implementation, 212, 218, 224, 229, 231, 233
ESG incorporation only, 39, 42, 44, 45
ESG integration, 12, 38–40, 42, 43, 45, 46, 67, 78
ESG investment choices, 65
ESG investment opportunities, 82
ESG lending practices, 123
ESG management system, 225, 226, 232
ESG metrics, 30, 32
ESG perspective, 219, 230, 233
ESG principles, 218, 220, 221
ESG ratings, 28, 30–33
ESG ratings reliability, 32
ESG regulations, 29, 33
ESG risk rating, 232
ESG risks, 10, 36, 39
ESG selection criteria, 75
ESG strategies, 220–223, 226
ethical and moral issues, 1
ethical banks, 116, 137–143
ethics, 2–4, 6, 11, 21
European SRI market, 37, 38, 45
European VC funds, 221, 222
EU sustainable finance taxonomy, 35
exclusion strategy growth, 13
exposure to climate risks, 110
expressive benefits, 22, 23, 26
extra-financial expectations, 24, 27

F

female entrepreneurs, 241–244
female VC investors, 219, 244
finance digitisation, 155
financial decision-making, 1
financial inclusion, 120, 138, 177, 191, 193, 196

financial partners, 131, 136
financial performance, 119, 122, 126–128, 136, 140
financial performance indicators, 225, 229
financial performance of SRI, 12, 17–19
financing instruments, 236, 241
fintech, 155, 184, 185, 188
fintech startup, 155, 156, 184, 185
five social principles, 69
forms of impact investing, 74
frameworks for impact assessment, 234
funding options, 240, 241
fund-matching, 184, 188, 190

G
GAVB network, 140, 141
GDP and bank leverage, 107
Global Reporting Initiative (GRI), 130, 131
global venture capital industry, 211
governance and social risks, 219, 224
governance assessment of a hedge fund, 87
governance issues, 108, 134
governance mechanisms, 118, 134, 140
governance pillar, 126, 131, 134
government interventions, 103
green crowdfunding, 195
green funds, 16, 20
green lending, 110

H
hedge fund strategies, 60, 85
herd behaviour, 163, 164, 168, 174
heterogeneity of the social screens, 19
higher level of deposits, 143
high financial performance, 65
holding period, 216, 233
holistic utility function, 22
homo economicus, 1
human resources, 176, 184

I
impact finance, 191
impact investing, 16, 38–40, 42, 43, 45
impact investment criteria in PE, 79
impact investment journey, 71
impediments to SRI, 29
increasing demand for ESG, 224
incumbents, 156, 160, 166, 185
individual investors, 12, 21, 27–29
informational cascade, 165, 168
information asymmetry(ies), 12, 24–28, 30, 160, 161, 164, 166, 168, 226–229, 243
institutional investors, 62–64, 66, 68, 70, 80, 86
institutional investors ESG criteria, 34, 42
integration of ESG factors, 64, 75
integrity, 137
investment approaches, 28, 38, 41
investment focus, 237
investment managers, 67, 85
investment period, 80, 81, 215–217
investment phase, 231
investment screening, 189
investment strategies, 13, 15, 28, 36, 38–42, 44, 46, 61, 64, 71, 72, 75–77, 79, 80, 85, 86
investor benefits, 63, 65, 66
investors approach, 63
Italian banking system, 105

L
lack of consistency, 31, 32
lack of transparency, 31

lending crowdfunding, 160, 167, 170, 181, 186, 195
liberalisation, 4
loan quality, 141
loans over total assets, 142, 143
loss of trust, 137

M

match-funding, 187, 190
modern alternative investments, 62
Modern Portfolio Theory, 17, 18
monitor ESG performance indicators, 233

N

negative screening, 13, 14, 19, 39
neutral performance of SRI, 19
nexus agency model, 24, 26–28
non-bankable/underbanked, 191
non-professional investors, 157, 178
norms-based approach, 15

O

open innovation processes, 171
operating models of social VC funds, 240

P

PE investment decision-making process, 76, 78
PE investments, 74, 81
PE investors, 60, 74, 76–80, 82–84
performance improvement, 61
portfolio activities, 77
positive and measurable impact, 72
positive screening, 14, 15, 19, 20
post-contractual opportunism, 25
pre-purchase crowdfunding, 169
Principles for Responsible Banking (PRB), 111, 112
private equity (PE), 60–62, 70, 72–75, 79, 80
profit and the costs of social responsibility, 10
profit maximisation, 6
property valuation tools, 91
pro-social lending, 181, 195
psychological benefits, 22

Q

quality of information, 227, 228

R

reduction of volatility, 63
reinforcement model, 163
relevant and robust metrics, 80
reputation-building theory, 227
resource exchange, 166
responsible investment, 64, 68, 71, 74–76, 87
retail investors, 30, 34, 37
reward-based crowdfunding, 159, 161, 164, 168–170, 174, 188, 192–194
riskiness of the decision, 8
risks of greenwashing, 112
role of impact investment, 70

S

satisfaction of social needs, 9
savings safeguarding, 113
separation model, 1–3, 6, 9–11, 21, 23
serial crowdfunding entrepreneurs, 169
shareholder activism, 86
shareholder resolutions only, 39, 42, 44

social capital, 162, 168, 169, 171, 177, 179, 190, 193
social communication, 129, 130
social enterprises, 158, 159, 189, 190, 192, 193, 196
social impact lending, 182
social inequality, 3
social legitimacy, 129, 133
social lending, 195
socially oriented investments, 194
socially responsible banking (SRB), 101, 108, 113, 117, 186, 189
socially responsible investing (SRI), 11, 12, 15, 17, 27
socially responsible investment choices, 21
socially sustainable business, 109
social-market activity, 230
social performance, 128–130
social pillar, 120, 131
social reporting, 176, 177
social reporting systems, 135
social return on investment (SROI), 235
the social role of bank, 101, 111, 113
social VC firms, 211, 234
social VC fund's strategic orientation, 236
the spectrum of capital, 73
SRFs and CSR, 122, 123
SRI mainstream, 28
SRI market growth, 29
SRI niche, 28
SRI orientation, 85
SRI strategies by countries, 40
stages of the investment process, 222, 231
stages of VC investments, 212, 213
stakeholder engagement model, 182
stakeholder governance model, 139
stakeholders interests, 30, 32
stakeholder theory, 2, 8, 18, 176, 182
standardisation of the information, 130
stereotypes and gender homophily, 243
suppliers, 118, 131, 136
supply of equity capital, 122
sustainability and social policy, 80
sustainability in the real estate sector, 91, 92
sustainability-themed investments, 13, 16, 39, 45
Sustainable Development Goals (SDGs), 33, 38, 179, 180
sustainable finance, 179
sustainable finance legislation, 36
sustainable oriented investors, 37
sustainable performance, 182
sustainable urban development, 89

T

taxonomy performance thresholds, 35
traditional alternative investments, 61, 62
triple-bottom-line logic, 10
twofold fiduciary mandate, 27
type and intensity of screens, 19
types of VC firms, 214

U

US market trends, 42
utility function, 8, 22, 24–26

V

value co-creation (VCC), 157, 171–176
value creation, 59, 67, 71
value creation streams, 67
value-in-use (ViU), 173, 174
VC-backed firms, 216, 219, 220, 224, 229, 241

VC firm's investment decision, 232
VC firm's perspective, 226
VC firms' fundraising, 223
VC fund business model, 214
VC fund manager, 221, 223
VC funds, 212, 214–217, 219, 224, 226, 231–237, 240–242, 244
VC funds' motivations, 221
VC gender gap, 212, 223, 241–244
VC industry, 212, 218, 220, 223, 224, 226, 242–244
VC industry performance, 224, 244
VC investment opportunities, 235, 242
VCs' success, 223
Vehicles for impact investments, 72
virtual environment, 161, 165
virtuous cycle between bank and company, 115, 117

W

waste management, 16, 40
wisdom of crowd, 157, 160, 165, 168, 171, 194
word-of-mouth (WOM), 162, 165, 166
world famous social VC funds, 234, 236, 237

The manufacturer's authorised representative in the EU is Springer Nature Customer Service Centre GmbH, Europaplatz 3, 69115 Heidelberg, Germany. If you have any concerns regarding our products, please contact ProductSafety@springernature.com

Printed and bound by CPI Group (UK) Ltd, Croydon, CR0 4YY
30/09/2026
02225702-0009